Advances in Atom and Single Molecule Machines

More information about this series at http://www.springer.com/series/10425

Marek Kolmer · Christian Joachim
Editors

On-Surface Atomic Wires and Logic Gates

Updated in 2016 Proceedings of the International Workshop on Atomic Wires, Krakow, September 2014

Springer

Editors
Marek Kolmer
Faculty of Physics, Astronomy
and Applied Computer Science,
Centre for Nanometer-Scale Science
and Advanced Materials, NANOSAM
Jagiellonian University
Kraków
Poland

Christian Joachim
CEMES-CNRS
Toulouse
France

ISSN 2193-9691 ISSN 2193-9705 (electronic)
Advances in Atom and Single Molecule Machines
ISBN 978-3-319-51846-6 ISBN 978-3-319-51847-3 (eBook)
DOI 10.1007/978-3-319-51847-3

Library of Congress Control Number: 2016963176

Printed on acid-free paper

This Springer imprint is published by Springer Nature
The registered company is Springer International Publishing AG
The registered company address is: Gewerbestrasse 11, 6330 Cham, Switzerland

Preface

Driving the ultimate miniaturization of electronic circuits down to the atomic scale motivates the investigations of novel calculator and memory architectures and concepts, which involve a supported atomic structure or even a single molecule, to act as an independent calculating unit. This was the motivation of the third volume of the Springer Series "Advances in Atom and Single Molecule Machines" entitled "Architecture and Design of Molecule Logic Gates and Atom Circuits", to bring together the different theoretical architectures for a molecular or atomic systems to perform an elementary Boolean logic function at the quantum realm. For this new "On-Surface Atomic Wires and Logic Gates" Volume 10 of the series, we go a step further in this burgeoning field of atomic scale wires and circuits. To be more practical, this volume 10 is focussing on the hydrogenated (001) surfaces of silicon and germanium at a supporting surface to construct those calculating circuits atom by atom. On hydrogenated (001) surfaces, atom-by-atom extraction of single hydrogen can construct unique pre-designed and atomically perfect dangling bonds (DBs) structures with well identifiable states in the electronic band gap of those surfaces.

Starting from the description of two experimental processes to fabricate Si(100)H and Ge(0001)H surfaces, Chapters "Surface Hydrogenation of the Si(100)-2x1 and Electronic Properties of Silicon Dangling Bonds on the Si(100):H Surfaces"–"Si(100):H and Ge(100):H Dimer Rows Contrast Inversion in Low-Temperature Scanning Tunnelling Microscope Images" present the basic and measured electronic properties of DBs. This includes the standard laboratory ultrahigh vacuum (UHV) preparation protocols of passivated surfaces (Chapters "Surface Hydrogenation of the Si(100)-2x1 and Electronic Properties of Silicon Dangling Bonds on the Si(100):H Surfaces" and "Atomic Wires on Ge(001):H Surface"). In Chapter "Nanopackaging of Si(100)H Wafer for Atomic Scale Investigations", a more microelectronic industry like fabrication process is described with its nanopacking possibilities. Those chapters show also experimental strategies for atomic scale DB structures construction with the use of the scanning tunnelling microscopy (STM) vertical manipulation protocol. These results come together with a detailed analysis of DB electronic properties, which are determined experimentally by scanning tunnelling spectroscopy (STS) supported by advanced theoretical modelling methods. The combination of experimental and theoretical methodologies allows the identification of some interesting phenomena such as the formation of dispersive band structures during the DB wire atom-by-atom construction (Chapter "Atomic Wires on Ge(001):H Surface") or the counter intuitive half-row shift of STM constant current contrast observed for both Si(001H) and Ge(001)H substrates along the SiH dimer rows (Chapter "Si(100):H and Ge(100):H Dimer Rows Contrast Inversion in Low-Temperature Scanning Tunnelling Microscope Images").

In Chapters "Band Engineering of Dangling-Bond Wires on the Si(100)H Surface" and "Band Engineering of the Si(001):H Surface by Doping with P and B Atoms", the quantum engineering of the electronic structure of single and long DB wires is described. Chapter "Band Engineering of Dangling-Bond Wires on the Si(100)H Surface" details the geometrical arrangements of the DBs along single DB wires with the prospect to minimize the electronic energy band gap of those wires and increase the dispersion of their valence and conduction bands. This can also be obtained by a very specific surface doping with boron or phosphorus atoms well positioned along those long DB wires as described in Chapter "Band Engineering of the Si(001):H Surface by Doping with P and B Atoms".

Finally, different designs of simple Boolean logic circuits using DBs complemented by molecular latches are shown in next four Chapters. Starting from the experimental and theoretical analysis of atom-by-atom single DB wire construction and its corresponding STS characterization (Chapter "Electronic Properties of a Single Dangling Bond and of Dangling Bond Wires on a Si(001):H Surface"), Boolean logic circuits are described including purely atomic DB structures (Chapter "Quantum Hamiltonian Computing (QHC) Logic Gates") or with when necessary and for more realistic logical input, molecular latches adsorbed on hydrogenated surfaces (Chapters "The Design of a Surface Atomic Scale Logic Gate with Molecular Latch Inputs" and "Molecule Latches in Atomic Scale Surface Logic Gates Constructed on Si(100)H"). First experimental implantations of quantum Hamiltonian logic gates are also described. Volume 10 ends with Chapter "Complex Atomic-Scale Surface Electronic Circuit's Simulator Including the Pads and the Supporting Surface" describing what will be in the future an atomic scale circuit simulator expected to guide the construction of very complex DB logic circuits on hydrogenated surfaces.

Similar to the others volumes of the series, contributions included in this volume mainly result from a workshop organized by the AtMol and PAMS Integrated European projects. For this volume, it was the "International Workshop on Atomic Wires" held in Krakow from 10 September 2014 to 12 September 2014. Selected contributions were then completed by new results obtained in 2015 up to mid-2016 to finalize a volume in a comprehensive form.

We would like to thank the European Commission's ICT-FET projects AtMol and PAMS for their financial support. We are also grateful to all of the authors who have contributed their work to this volume and Mlle Marie Hervé for her help in formatting this volume and many others volumes of the series.

Kraków, Poland
Toulouse, France
October 2016

Marek Kolmer
Christian Joachim

Contents

About the Editors

Marek Kolmer obtained his Ph.D. degree in experimental physics in 2014 from the Jagiellonian University in Kraków (Poland) under the supervision of Prof. Marek Szymoński. After 6 months attachment to the NanoSciences Group at CEMES (CNRS) in Toulouse, he came back to Krakow, where he is now a researcher at the Faculty of Physics, Astronomy and Applied Computer Science, Jagiellonian University. His research interests are related to fabrication and experimental verification of functional properties of molecular and atomic structures on semiconducting surfaces, i.e. hydrogenated surfaces of silicon and germanium. Despite an early stage of his scientific career, he is the author of more than 20 publications. His current scientific achievements were also rewarded by a few national awards including ETIUDA grant from the Polish National Science Center (2013) and START scholarships from the Foundation of Polish Science (2014, 2015).

Christian Joachim is Director of Research at the Nanoscience group in the Pico-Lab CEMES/CNRS and adjunct Professor of Quantum Physics at ISAE Toulouse. He was A*STAR VIP Atom Tech in Singapore (2005–2014) and is the head of the WPI MANA-NIMS satellite in Toulouse since 2008. He coordinated the Integrated European projects “Bottom-up Nanomachines”, “Pico-Inside” and “AtMol” (2011–2014) whose objective was to prepare the construct of the first ever molecular chip. He is the author of more than 270 scientific publications (h = 55) and had presented over 360 invited talks on electron transfer through a molecule, STM and atomic force microscopy (AFM) image calculations, tunnel transport through a molecule, single molecule logic gate, atomic scale circuits, nanolithography, atomic scale electronics interconnects and single-molecule machines. His book “Nanosciences, the invisible revolution” (Le Seuil 2008; World Scientific 2009) is giving the history of nanosciences and its political drawbacks to a general public. He was awarded the IBM France Prize (1991), the Feynman Prize (1997), the CNRS Silver Medal in Chemistry (2001), the Feynman Prize (2005) and a Guinness book entry (2011) for the smallest ever functioning nanogear, 1.2 nm in diameter.

Surface Hydrogenation of the Si(100)-2×1 and Electronic Properties of Silicon Dangling Bonds on the Si(100):H Surfaces

Damien Riedel

Abstract In this chapter, we give a short review about the methods that are commonly used to passivate the Si(100) surface with hydrogen atoms. The wet technique is discussed in terms of surface pollution and surface roughness. A basic recipe is given. A second part is devoted to the methods commonly used with vacuum techniques. A discussion is done on the hydrogenation parameters to improve the surface quality at the atomic scale, in particular the one that concerns the formation of various phases such as the 3×1 and dihydride. A third method is also detailed and permits the surface hydrogenation at the atomic scale allowing to design some patches on the Si(100) surface of a few nanometer.

1 Introduction

The control of the surface properties, whether it deals with metallic or semiconductor surfaces, is of major interest in many domains including chemistry [1, 2], optics [3], and nanoscience [4]. For catalysis [5], self-assembled layers, or molecular superstructure [6] as well as epitaxial reactions [7] and electronics [8], the interaction of the surface with the adsorbates has a major role, especially when semiconductor surfaces are of interest. In this context, the passivation of surfaces is of crucial importance to allow the tuning of their chemical and physical properties.

D. Riedel (✉)
CNRS, Institut des Sciences Moléculaires d'Orsay (ISMO), Université Paris-Sud,
Université Paris-Saclay, 91405 Orsay, France
e-mail: damien.riedel@u-psud.fr

M. Kolmer and C. Joachim (eds.), *On-Surface Atomic Wires and Logic Gates*,
Advances in Atom and Single Molecule Machines,
DOI 10.1007/978-3-319-51847-3_1

One of the most famous examples in this topic is the passivation of the silicon surface with hydrogen because of its general interest in the fabrication of microelectronic components [3]. The use of the scanning tunneling microscopy (STM) on the bare silicon, firstly, and then on the hydrogenated silicon surface had a central part in for the understanding of their its electronic properties at the atomic scale. In particular, many studies have been devoted to the analysis of the hydrogen reaction on surfaces [9, 10]. In this chapter, we will review the hydrogenation methods used to obtain two types of passivated surfaces: (i) in air, mainly for large-scale purposes and (ii) in ultrahigh vacuum (UHV) to have atomically flat hydrogenated surfaces. The first technique is a wet method using chemical treatment in diluted acids. The second method, the mostly used, is based on the interaction of atomic H and excited H_2^* in UHV with the bare surface kept at a given temperature. A third method, more recently elaborated, uses the STM tip to locally passivate the silicon bare surface via ad hoc electronic excitations [11]. Then, we will discuss some of the side effects of the hydrogenation process such as the formation of dihydride structures or the influence of the dopant concentration and its interaction with the interstitial H atom in the silicon subsurface. The second part of this chapter will be devoted to show some investigations of the Si(100):H surface and in particular on the electronic properties related to the silicon dangling bonds (Si-DB).

2 Hydrogen Passivation Method of the Si(100)-2×1 Surface

2.1 Wet Etching Methods

The wet H-passivation methods have been initially elaborated from the techniques used in microelectronic processes in order to prepare silicon surfaces for controllable thin SiO_2 epitaxy. Although these methods can bring some inconvenients such as the formation of additional surface reconstructions or subsurface contaminations, it has been successfully used to prepare atomically flat silicon Si(111) samples [12, 13]. The etching solution that allows the preparation of the Si(100):H or Si(111):H surfaces in ambient conditions is usually prepared with HF diluted in a solvent (usually 1:100 parts). The choice of the solvent is crucial. Indeed, it has been showed that the use of water solvent usually induces oxygen surface pollutions such as oxygen oxidizing and adsorption. Light alcohol is usually preferred as it has more affinities for Si surfaces and helps in forming H-terminated surfaces due to their ease of evaporation. Hence, the use of methanol, ethanol, or propanol as a diluent has often been tested to obtain clean H-passivated Si(100) surfaces in particular because these solvents are easier to evaporate and also help to avoid the haze observed during the HF processing.

This surface preparation process is most often preceded by the so-called RCA-type cleaning established in 1965 by the Radio Corporation of America. This cleaning process defines 3 main steps. The first step called SC-1 (standard cleaning-1) is used for organic + particle cleaning. The SC-1 solution is made of 5 parts of deionized water, 1 part of ammonium hydroxide (NH_4OH), and 1 part of hydrogen peroxide (H_2O_2). The SC-2 solution which purpose is to remove ionic metal contaminants is prepared with 5 parts of deionized water, 1 part of hydrochloric acid (HCl), and 1 part of aqueous hydrogen peroxide (H_2O_2).

The H-passivated surface is then generally obtained by proceeding as follows: (i) dip the silicon wafer in the SC-1 solution at 80 °C for 10 min; (ii) rinse in deionized water (19 MΩ) for 5 min and dry with pure N_2; (iii) dip in HF:H_20 (1:100) solution for 2 min [14]; (iv) rinse in deionized water (19 MΩ) for 5 min and dry with pure N_2; (v) dip in SC-2 solution at 80 °C for 10 min; and (vi) dip in HF:alcohol solution for 2 min at room temperature.

Some research groups have been able to optimize these methods in order to obtain very flat Si(111)-(1×1):H surfaces that could be UHV compatible. In the work of Kato et al. [12], it is explained that there are several important issues to succeed in preparing ultraclean Si(111)-(1×1):H surfaces. One of them concerns the etchant used NH_4F instead of HF which is shown to give better results for atomically flat and smooth surfaces. They could investigate the atomic structure of the treated Si(111)-(1×1):H surface with STM running at room temperature and investigate the low concentration of contaminants with STS spectra. Additionally, they could obtain atomically resolved images with a corrugation lower than 0.04 nm [12].

The preparation of the Si(100):H with wet etching methods is mainly used as a preliminary process to prepare clean and atomically flat Si(100)-2×1 surfaces to avoid the use of high temperature. Indeed, with the Si(100) surface, the wet etching methods can hardly produce atomically flat surfaces that can be used with UHV STM [15]. Additionally, the obtained terraces are usually of small size compared to other UHV methods. Another drawback related to Si(100):H surfaces prepared with wet etching is related to the formation of pyramids that reveals Si (111) facets on the initial Si(100) surface. This effect is principally related to the HF concentration of the etching solution. Recent work on this subject has been performed on Ge(100) surfaces [16].

In a more general purpose, the preparation of the Si(100):H surfaces at the atomic scale is mainly related to its electronic decoupling properties and exploits surface point defects such as Si-DB when used as quantum dots. In this framework, it is crucial to produce atomically flat surfaces of Si(100):H made of large terraces with the minimum of defects, whereas the Si(111):H surface is generally less studied due to its lack of interests in the atomic scale because of the unstable dihydride (1×1) phase.

2.2 In Situ Preparation

The preparation of an atomically flat hydrogenated Si(100) surface that are UHV compatible is mainly based on the methods reviewed by Boland et al. [17]. Since then, very few developments have been realized in finding other passivation methods [18, 11]. Nevertheless, regular improvements in this method related to the adjustment of the temperature of the silicon sample or the filament that is used to crack the H_2 molecules are performed. As we will see in this chapter, the size of the passivated terraces, the number of defects (missing dimers, residual dangling bonds, pollution), and the formation of monohydride or dihydride phases on the Si(100):H surface or the migration of subsurface dopants in the bulk silicon can be controlled by playing on the sample preparation parameters.

Hydrogenation of the Si(100) surface has several important impacts on the silicon as it modifies the chemical reactivity and the electrical conductivity of the surface and the subsurface. Hydrogenated surfaces are passivated and hence are much less reactive than the bare Si(100) surface. The physical properties of the silicon are also modified because several surface states can be erased as well as bulk states because hydrogen atoms can penetrate into its subsurface and reacts with dopants. One of the main effects of the hydrogenation process on the bare Si(100) surface properties is to enlarge its surface band gap energy [19]. Simultaneously, surface hydrogenation methods such as the one we have seen in the previous Sect. 1 can help to remove defects and thus improve interfaces in microelectronic devices.

Several review articles have been published over the last 15 years on the interaction of hydrogen with surfaces. it deals with hydrogen either on metal surfaces [20] or on the semiconductors [21]. For semiconductors, one of the well-known review articles for the hydrogenated surface preparation is the one of Boland's [17]. Another good report article written by Oura et al. [22] treats about the reactivity of hydrogen with silicon when adsorption of metals is made. Here, we will mainly describe the most common method that is still used nowadays to prepare the hydrogenated surface in order to obtain UHV compatible atomically flat 2×1 passivated Si(100) surface. The main principle is to create reacting H_2 molecule in an excited state or H atoms that are exposed to bare Si(100)-2×1 surface. The methods used to obtain excited H_2 or single H atoms can vary. Some experiments use a gas H_2 cracker tool to expose the Si(100) surface to a flux of H atoms (see Fig. 1a) [23]. Initially employed to clean surfaces, the cracking method is not so often derived for the hydrogenation process of silicon surfaces mainly because of its price. An alternative method, most commonly employed, uses a hot tungsten filament to replace the cracking process of the H_2 (Fig. 1b). In the same time, the silicon sample is heated to increase the reaction rate. We will see that in this case, the sample temperature adjustment during the passivation process is very important in order to control the hydrogenation process as it is ruled by an adsorption–desorption cycle. Hence, an important parameter that allows to adjust

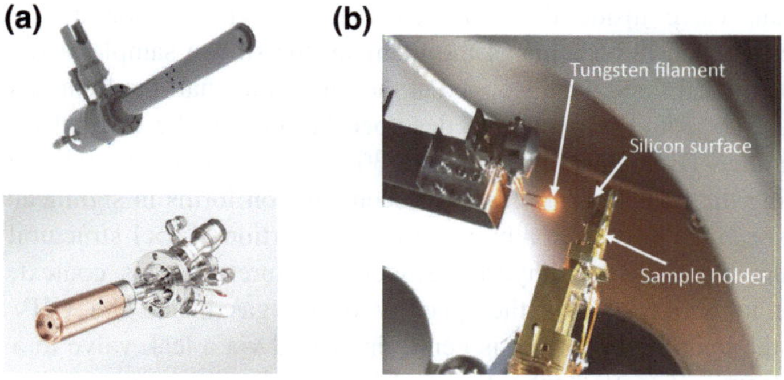

Fig. 1 **a** Example of gas cracker used for the surface passivation with hydrogen (courtesy from RIBER and MBE). **b** The heated tungsten filament creates H atoms to react with the silicon surface placed in front

the passivation is the hydrogen exposure. It is expressed as the product of the pressure and the time and is given in Langmuir (L) where 1 L corresponds to a pressure of 1×10^{-6} Torr for the duration of 1 s. One can also use the monolayer unit (ML) where 1 ML corresponds to the occupation of every surface site by an adsorbed atom. For example, there are 6.8×10^{14} atoms cm^{-2} at the first atomic layer of the Si surface.

To prepare an hydrogenated surface, the first step is to clean the Si(100) surface. This is done firstly to remove the native oxide layer that is formed on top of the surface. This procedure is done by heating the silicon sample up to 900–1000 °C for a short period of 10–30 s. It is repeated several times until the sample is completely exempt from surface pollution. Then, it is needed to proceed with the reconstruction of the silicon surface. For this, the sample is rapidly heated to $\sim$1100 °C for a very short time (few seconds), and then, its temperature decreased to $\sim$950 °C. The decay in temperature from 950 to 650 °C is then performed very slowly in $\sim$120 s. These two steps, usually called 'flashing' [24], are repeated several times until a good surface reconstruction is obtained and can be further checked via LEED techniques.

Once the clean Si(100)-2×1 surface is prepared, the sample is place in front of a tungsten filament (at about 2–10 cm). During this process, the sample is kept at a temperature of $\sim$360 to 395 °C. The tungsten filament is heated to a temperature that can vary from 1500 to 2000 °C according to various articles [21]. This seemingly large range of temperature leads the experimentalist to choose its own value and adjust it. It is however important to note that the more the tungsten filament is hot, the more it will affect the initial silicon substrate temperature due to the radiation when located near the W filament. When the filament and the sample are located correctly, the hydrogen molecule can be introduced inside the UHV chamber. The quality of the H_2 gas is crucial. It is better to choose high-grade 5.6 H_2 gas and to use a liquid nitrogen trap along the horse that connects the H_2 bottle to the leak valve inside the UHV chamber. Another important aspect is the sequence at which the heating of the filament, the silicon sample, and the H_2 flux are stopped. If all three were stopped at the same time, half the hydrogen adsorbed on the surface would be thermally desorbed because of the thermal inertia of the sample holder. Differently, if more than 30 s elapses between cooling the sample and cooling the hot filament, the 3×1 reconstruction forms in significant quantity. We will see in the next section how a strong proportion of 3×1 structural phase can lead to the formation of dihydride silicon structures. In this context, it is also important to correctly dose the quantity of H_2 gas inside the UHV chamber. Usually, the hydrogen pressure is gently increased via a leak valve to a value that can vary from $\sim 2 \times 10^{-8}$ mbar to 7×10^{-6} mbar for the duration varying from 60 to 1200 s. This corresponds to a range of exposure of 3.6–1400 L. In practice, the more the coverage is high, the more the 3×1 phase is favored which is also combined with the sequence at which is stopped the hydrogenation process as noticed previously. Another important point is related to the remaining dangling bond after the hydrogenation process. It is very difficult to obtain Si(100):H surface with no remaining silicon dangling bonds because their formation is a multifactor process. Additionally, although the hydrogenation process is known to slightly etch the silicon surface, the size of the obtained terraces after the hydrogenation procedure is mainly related on how the bare surface has been prepared previously. In summary, the hydrogenation procedure can be run as follows:

(a) Prepare in UHV a clean Si(100)-2×1 surface by repeating heating cycles 1100 °C. When ready, the sample temperature should be kept at a value around $\sim$380 °C. This can be adjusted via the resistive heating.
(b) Place the silicon sample in front of the tungsten filament (2–10 cm) that is heated at an intermediate temperature ($\sim$900 °C).
(c) Proceed with the dosing of the H_2 gas inside the vacuum chamber introducing it via a leak valve and increasing the pressure up to $\sim 1 \times 10^{-6}$ Torr.
(d) Increase the current that heat the tungsten filament to a calibrated value to obtain a temperature of $\sim$1750 °C (typically 3–4 A).
(e) Start the stopwatch and keep the parameters stable for the duration of −1200 s.
(f) Stop first the tungsten filament heating and after few seconds reduce the temperature of the sample to $\sim$100 °C.
(g) Stop the exposure of the H_2 gas by closing the leak valve.
(h) Move the sample holder away from its location near the tungsten filament.
(i) Wait several minutes for the remaining H_2 to be pumped away from the UHV chamber.
(j) Set the temperature of the obtained hydrogen sample to the desired one and the surface studied.

This method is an example of what can be done to obtain a good quality Si(100):H surface. All the cited parameters may be slightly modified according to the user needs.

2.3 Side Effects of the Hydrogenation: Monohydride and Dihydride Phases

As cited above, the preparation of the silicon sample during the hydrogenation process is delicate to control. In particular, the temperature of the sample at which the hydrogenation is exposed can imply the production of various silicon hydride phases. The work of Bellec et al. illustrates perfectly well these phenomena [25]. In this work, the sample is prepared as follows: Si(100) samples are *n*-type As-doped, with a resistivity of 5 mΩ cm, and *p*-type *B*-doped, with a resistivity of 6 mΩ cm. After preparing clean Si(00)-2×1 surfaces under UHV base pressure 1.10^{-10} Torr, hydrogenation of the clean Si(100) surface is performed, as previously reported, with the sample kept at 650 K [17, 21]. After hydrogenation, the sample is cooled down to 12 K and transferred to the STM chamber. Figures 2a–c illustrates the

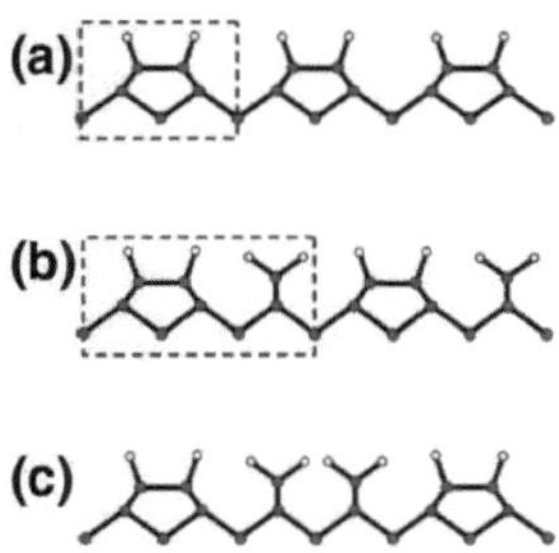

(d)

Sample	2×1	3×1	bow-tie	other
n type (As doped)	66.8 ± 0.5%	23.0 ± 0.5%	5.3 ± 0.2%	4.9 ± 0.2%
p type (*B* doped)	79.8 ± 0.3%	9.3 ± 0.2%	3.9 ± 0.2%	7.0 ± 0.2%

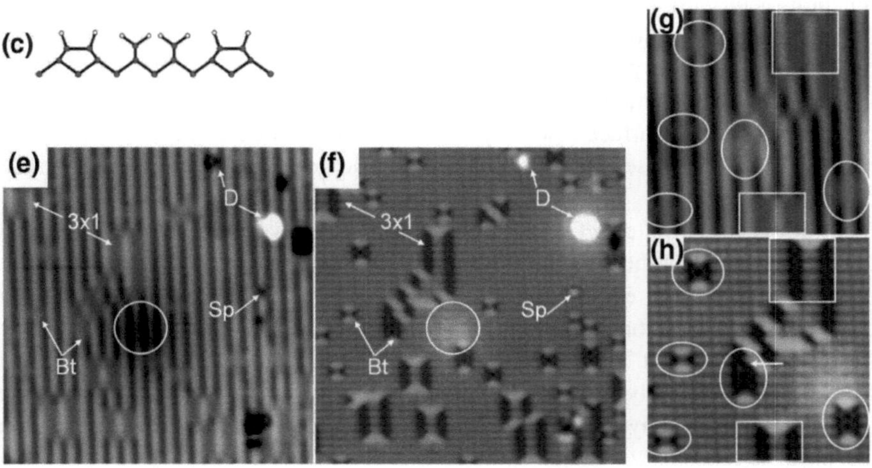

Fig. 2 **a–c** cross-sectional sketches of the Si(100):H surface showing the 2×1, 3×1, and dihydride structures, respectively. **d** Table indicating the proportion of each surface reconstruction phase observed on the n-type or p-type-doped Si(100):H surface. **e** and **f** 170 × 170 Å^2 STM topographies of a p-doped Si(100)-2×1:H surface at 5 K for **e** filled states (Vs = −2.5 V and I = 110 pA) and **f** empty states (Vs = 1.7 V and I = 110 pA). Bt stands for the bow-tie structures, Sp for the split dimer structures, and D for unknown defects. The *white circle* indicates the presence of a subsurface dopant. **g** and **h** STM topographies of the p-type Si(100):H surface (83×83 Å^2), **g** sample bias V_S = −2.5 V and tunnel current I = 110 pA, and **h** sample bias Vs = 1.7 V and tunnel current I = 110 pA. *Rectangles* and *circles* show the 3×1 and the dihydride dimer reconstructions

cross-sectional view of various reconstructions that can be obtained on the Si(100):H surface after the hydrogenation process.

The Si(100)-2×1:H (Fig. 2a) can be obtained in the majority of the cases (66–79% of the surface area), while the Si(100)-3×1:H is only observed in 9–23% (see Fig. 2b). In these conditions, the so-called bow-tie (BT) structure has been often wrongly attributed to dopant segregation on the surface forming a pair of P atoms within the silicon dimer row [26]. In fact, the formation of 3×1 leads to the creation of bow-tie structures that are made of two consecutive dihydride silicon atoms (Fig. 2c). This type of structure is observed when the sample temperature is relatively low (650 K or 377 °C in our case) and when the dynamics of the hydrogenation process is relatively slow to allow the diffusion of 3×1 phases from the step edges toward the terraces. As we can see in the series of two STM pictures of the hydrogenated surface (Fig. 2e and f), the 3×1 and the BT structure can be clearly observed, in particular when the unoccupied states are probed (Fig. 2f). Additionally, the presence of subsurface dopant atoms can also be detected via the observation of a large fuzzy depletion of charge density in the STM topography (see the white circle in Fig. 2e and f). The presence of split dimer (SP) that can be observed either at room or at low temperatures cannot be ascribed to dihydride dimers. This type of defect is not well understood yet and need further investigations [25]. The formation of dihydride dimers can be further understood when one look at the series of STM topographies as illustrated in Fig. 2g and h in which the junction of two lines of 3×1 structures join to form a dihydride line of 3 silicon dimers' length. The formation of these surface defects should be avoided if one needs to exploit the periodicity and homogeneity of the Si(100)-2×1:H phase. However, in some particular cases, the control of the formation of dihydride dimers might be useful for surface reactivity. These structures are initially less stable than the Si(100)-2×1 phase and can lead to specific reactions during adsorption of devoted molecules.

More recently, the effect of the dopant depletion have been studied during the initial preparation of the bare Si(100) sample [27]. In particular, what can be probed in the work of Labidi et al. is the influence of charge transport through individual dangling bonds as a function of the dopant depletion in the subsurface. The diffusion of the dopant away from the subsurface is produced during the flash annealing at a relatively high temperature of ~1250 °C for a time as short as 2 s. In these conditions, the hydrogenation process is not directly affected since the authors can produce relatively good quality Si(100)-2×1 surface. In this case, the electronic properties of Si-DB can be strongly affected if the sample preparation is not correctly controlled. Here, depletion might also be exploited to tune the ratio of electronic transport between the silicon bulk to the Si-DB and the one from the DB to the STM tip. In such a case, the precise control of a homogeneous dopant depletion in the subsurface would be tricky to monitor at the atomic scale. Hence, electronic properties of point defects such as the Si-DB and electronic interactions of adsorbed molecules with the Si(100) surface may, in this case, depend on their position on the surface.

2.4 Atomic-Scale Control of the Hydrogenation

As explained in the section above, the hydrogenation process of semiconductor surfaces has very little changed since the work of Boland et al. [17]. However, a recent progress has been achieved, showing that it is possible to locally passivate the surface of a Si(100) surface with H atoms. This effect is observed when the thermal shields of a low-temperature (9 K) STM is initially filled with pure molecular hydrogen (H_2) at a pressure of 1.0×10^{-6} Torr for 2 min. The STM tungsten tips employed are carefully prepared by chemical etching and cleaned by electron bombardment heating to remove their oxide layer. The silicon samples used during these experiments are n-type As-doped Si(100) samples ($\rho = 0.005\ \Omega$ cm) cleaned with a succession of resistive heating periods as previously described in the previous section. Then, the introduction of H_2 is performed, and STM topographies of the bare silicon surface can still be observed as shown in Fig. 3a.

The surface has a very weak density of defects which allows to find relatively large areas of the bare Si(100)-2×1 terraces. The STM tip is then located at a given position (the red dot in Fig. 3a), and the surface voltage is switched from the

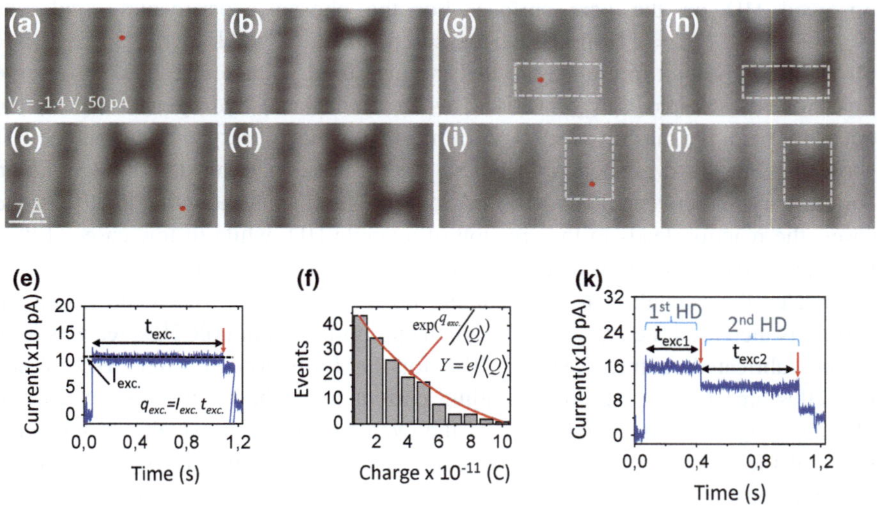

Fig. 3 **a–d** (3×1.5 nm^2) STM topographies (−1.4 V, 50 pA) of the bare Si(100) surface. **a** and **b** Before and after an electronically induced reaction (see *red dot* location at $Vs = +1.4$ V, 100 pA); and **c** and **d** before and after a second electronically induced reaction. **e** Typical current trace recorded during the excitation pulse **b** and **d**. **f** Distribution of the measured electronic charge ($I_{exc.} \times t_{exc.}$) required to induce the surface reaction observed in **b** and **d**. **g-j** (3×1.5-nm^2) STM topographies (−1.4 V, 50 pA) of the bare Si(100) surface. **g** and **h** Before and after an electronically induced reaction (at $Vs = +1.4$ V, 200 pA) on two successive horizontal dimers; and **i** and **j** before and after an electronically induced reaction (at $Vs = +1.4$ V, 200 pA) on two successive vertical dimers. The *red dot* indicates where the tip is placed during the excitation, and the *white dashed rectangles* are landmarks that locate the apparition of the HDs. **e** A typical current trace recorded during the electronically induced reaction in **h** or **j**

scanning conditions at negative bias ($V_s = -1.4$ V) to a positive bias $V_s = +1.4$ V. During this excitation procedure, the feedback loop of the STM is switched off and the tunnel current is recorded whereby the height of the STM tip is kept constant. Note that the tip height can initially be changed to adjust the current value. A typical tunnel current trace recorded during the excitation procedure is reported in Fig. 3e. During the time $t_{exc,}$ the tunnel current $I_{exc.}$ flows through the STM junction and then drops suddenly before the end of the current trace (red arrow in Fig. 3e). A second topography of the same area shows a short dark line centered on the dimer row and located beneath the tip apex position chosen for the excitation (Fig. 3b). This dark feature corresponds to a fully hydrogenated (HD) silicon dimer. Note that the creation of a HD during the excitation process is at the origin of the current drop detected in the current trace in Fig. 3e. The same procedure can be repeated at various positions, randomly chosen, to produce similar dark protrusions on the Si surface (Fig. 3c and d). For each excitation resulting in the appearance of a single HD, the charge $q_{exc.} = t_{exc.}\ I_{exc.}$ is calculated and its distribution over a large number of events is shown in Fig. 3f from which we can deduce an average charge $\langle Q \rangle$. The exponential decay of this distribution indicates that the observed processes are independent, i.e., induced by the tunnel electrons [28]. From this distribution, we can deduce an average quantum yield ($Y = e/\langle Q \rangle$) required to fabricate the observed HD on the bare surface. For the chosen experimental conditions ($V_s = +1.4$ V, $I_{exc.} = 110$ pA), the measured quantum yield is $Y = 2.8 \pm 0.7 \times 10^{-9}$. It is important to emphasize that the measured voltage threshold to observe this process is found at $V_s = +1.2$ V, whereas the local hydrogenation does not occur for negative bias up to -3.0 V. In most cases, the observed feature is the HD shown in Fig. 3a–d created right below the tip apex location. For the excitation current varying between 50 and 100 pA, our measurements show that in 87% of the cases, the reaction leads to the creation of a single HD, while in few cases (13%), two successive HDs can be created as shown in Fig. 3g–j (the red dot in Fig. 3g and i indicates the excitation location of the tip in each case). When the excitation current is increased (i.e., $I_{exc.} \sim 200$ pA), the proportion of double HDs reaches 60% of the fabricated features leaving 40% for the formation of single HD. When two HDs are created, the corresponding current trace shows two distinct current plateaus (Fig. 3k) which indicate that the events occur successively. A statistical study of the process related to the creation of two consecutive HDs is performed, and the ensuing average quantum yields are $Y_{HD1} = 2.7 \pm 0.9 \times 10^{-9}$ and $Y_{HD2} = 2.2 \pm 0.7 \times 10^{-9}$ for the creation of the first and the second HD, respectively. These quantum yields are very similar to the yield needed to create a single HD ($Y = 2.8 \pm 0.7 \times 10^{-9}$) which means that the probability to electronically induce a HD on the silicon surface is constant when the excitation current increases and thus arises from a single electron process [29]. Note that the creation of double HDs is also very local as it is observed either on two adjacent silicon dimer rows (Fig. 3g and h) or on two adjacent silicon dimers of the same dimer row (Fig. 3i and j). Our experimental findings suggest that the observed electronically induced process resembles a local hydrogenation of a silicon dimer via an *intra-dimer* dissociative adsorption [30]. Hence, we have performed STM topographies of the same scanning area of a bare silicon surface initially at a surface voltage ($V_s = -1.4$ V) that does not activate the surface reaction (Fig. 4a) and, subsequently, on the same area at a positive surface voltage ($V_s = +1.4$ V, I = 200 pA) to induce the fabrication of HD features (Fig. 4b). Following this procedure, a third STM topography at a larger scale is performed ($V_s = -1.4$ V) in which a dark square of about the size of the previous STM topographies appears (Fig. 4c). The comparison of dI/dV measurements performed on the bare surface and the dark area shows that the unoccupied π* surface states disappear on the created dark zone and show a surface gap of ~2 eV (Fig. 4d). These differences are characteristic of the electronic properties of hydrogenated surfaces as previously reported [31]. What could be the mechanism lying at the origin of the observed electronically induced local surface hydrogenation?

Our experimental results show that the hydrogenation occurs only on silicon dimers, via an *intra-dimer* hydrogenation process, involving the dissociation of H_2 molecules which origin must be precisely investigated. During the gas introduction in the thermal shields of the microscope, the cold STM (~9 K) is exposed to 120 Langmuir of H_2 (1.0×10^{-6} Torr for 120 s). In these conditions, H_2 molecules are rapidly condensed on each parts of the STM, leading to a negligible concentration

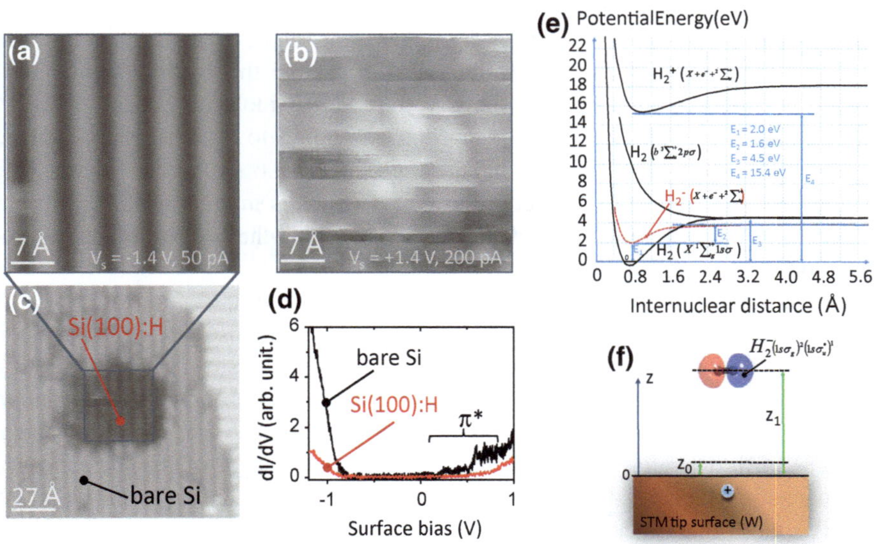

Fig. 4 **a** and **b** STM topographies (4.6 × 4.6-nm²) of the exact same surface area of the bare Si (100)-2×1 surface at −1.4 V, 50 pA and +1.4 V, 200 pA, respectively (scan speed of ~1000 Å/s for (b)). **c** The (14.6 × 14.6-nm2) STM topography (−1.4 V, 50 pA) of the bare Si(100) surface centered at the same location as in (b). **d** Averaged dI/dV spectroscopy curves acquired on the bare Si(100) (*dark curve*) and on the reacted area (*red curve*). **e** Potential energy curves of H_2 in gas phase extracted from [T.E. Sharp, Atomic data, **2**, 119 (1971)] **f** Sketch of an adsorbed H_2^- molecule on a metal surface showing the variables z_1 and z_0 included in the image charge potential equation $V^{e^-}_{\text{surf.}} = \frac{-q^2}{4(z_1 - z_0)}$

of residual H_2 gas in the STM junction [32]. However, we clearly observe that the STM topographies of the silicon surface are similar before and after the introduction of molecular hydrogen which indicates that the Si surface is not covered with a layer of molecular hydrogen, in particular in the region of the STM junction. This is consistent with the fact that at 9 K, H_2 molecules stick more efficiently on the tungsten tip compared to the silicon surface [33–36]. Surprisingly, it has been reported that H_2 molecules initially dissociate on crystalline tungsten at 9 K with a sticking coefficient of $\sim$0.6 to reach a saturating coverage of $\sim$0.8 monolayer [37, 38]. At the second step, the incoming H_2 molecules are physisorbed on the passivated areas to form adlayers of molecular H_2[39]. The electronic interaction of the molecular H_2 adlayer with the tungsten surface is therefore similar to what is observed on noble metal surfaces on which H_2 dissociation does not occur [40]. These observations indicate that in our experiments, the tungsten tip is covered with molecular hydrogen and acts as a source of H_2 for the hydrogenation reaction. This implies that in order to induce the silicon hydrogenation, H_2 molecules must be desorbed from the STM tip and then readsorb on the silicon surface. It is rather difficult to estimate precisely the desorption energy of the H_2 molecules from an amorphous tungsten tip surface as various possible adsorption sites are possible. Nevertheless, the order of magnitude of this desorption energy is reported to be lower than a few hundreds of meV [36, 41, 42]. Yet, the energy threshold observed experimentally to produce the local hydrogenation of the silicon dimer is much higher ($\sim$1.2 eV) than the energy required to activate the desorption of physisorbed H_2 from the STM tip. Additionally, once the H_2 molecule has left the STM tip, the *intra-dimer* adsorption of H_2 on Si(100) requires to reach an energy barrier of $\sim$0.6 eV. The transfer of this amount of energy ($\sim$0.6 eV) to the H_2 molecule would imply numerous multielectron processes which is in contradiction with our experimental observations. Indeed, we have shown that the observed local hydrogenation is induced via a one-electron process.

These arguments rule out the possibility to induce the desorption of H_2 molecules from the STM tip combined with their subsequent re-adsorption on the silicon surface via elastic tunnel excitations. Our experimental data are strongly suggesting that an electron attachment on the H_2 molecule trigger the observed hydrogenation process similar to what has been reported for molecular H_2 desorption on metal surfaces [43]. Therefore, the mechanism being at the origin of the local surface hydrogenation involves the charging of the H_2 molecules adsorbed on the STM tip. The electronic quantum yield measured in our experiment represents, consequently, the probability per electron to create H_2^- anion on the tungsten tip. The energy barrier U required to create H_2^- from molecular H_2 in the gas phase is $\sim$2 eV [44]. On a metallic surface, this value can be reduced ($U_{corr.}$) due to the presence of the image charge potential of the anion [45]. An estimate of this potential can be simply performed if the charge of the anion is regarded as a point charge which leads to consider, in our case, the simple interaction of a single charge q with the tungsten surface (see Fig. 4e). Under these conditions, the image charge potential for a H_2^- anion located at z_I above the surface (see Fig. 4f) is

$U = -q^2/4(z_1 - z_0)\, V_{inter} = -q^2/4(z_1 - z_0)$, where the effective position of the metal surface is represented by z_0. Considering an average value of z_1 ~4–5 Å for physisorbed H_2, we can estimate V_{inter} in the range ~0.9–1.1 eV. This leads to an energy barrier $U_{corr.}$ to create H_2^- on the tungsten tip in the range ~0.9–1.1 eV which matches well our measured experimental threshold required to induce the local hydrogenation of a silicon dimer (~1.2 V). This result strongly indicates that the ensuing local hydrogenation process is indeed related to the initial creation of H_2^- anion on the STM tip.

When H_2^- is formed on the STM tip, the anion is rapidly attracted by the positively charged silicon surface (V_s = +1.4 V) while simultaneously repulsed by the negatively polarized STM tip. This effect initiates the desorption of the H_2^- molecule from the STM tip and subsequently its transport toward the silicon surface by the electrostatic field applied in the STM junction. The time of flight of the H_2^- anion in an electrostatic field of 1.4 V over a distance of 1 nm before impacting on the silicon surface can be estimated as ~300 ps. This relatively short flying time warrants the H_2^- anion to be rapidly interacting with a silicon dimer prior to its dissociation or neutralization because its lifetime exceeds microseconds [46]. The charging process of a H_2 molecule from the STM tip combined with the attraction of the produced H_2^- anion toward the surface completely explains the observed local passivation process with hydrogen. It also clarifies why this local hydrogenation reaction cannot be observed for negative biases at which the creation of H_2^- anions might also be possible. This is mainly due to the repulsion of the created H_2^- anions away from the (negatively charged) silicon surface combined with an increased probability for the H_2^- anions to lose its extra charge in the (positively charged) tungsten tip. Note that the ionization of the H_2 molecules (H_2^+) can be excluded as it will require an energy barrier higher than 15 eV which is unreachable with tunnel electrons from an STM.

3 Electronic Structure of a Silicon Dangling Bond and Dehydrogenated Dimer

The development of passivation methods with hydrogen of the silicon surface has often been motivated by fundamental studies of the hydrogenated surface properties. Already, the bare Si(100) surface, although deeply investigated since the past decades, can always surprise us when new electronic or surface structure properties can still be discovered [47, 48]. Indeed, at the nanoscale, the study of the Si(100):H surface properties and its use has regularly influenced the passivation methods with hydrogenation. Similarly, recent experiments show that the initial process used to prepare the bare silicon surface before the hydrogenation can play an important role in the final properties of the surface point defects such as dangling bonds [27]. Hence, the study of atomic-scale electronic properties of Si-DB is of major importance as it is completely related to the silicon bulk properties and in particular in the presence of a large dopant concentration.

In addition, the development of molecular electronics requires specific substrates on which molecules can keep their intrinsic properties and have a good planar electronic contact as well. For this purpose, the Si(100):H surface appears as one of the best candidate because of its ability to electronically decouple adsorbed molecules from the silicon substrate [49]. Moreover, this surface can be considered as a flexible medium as it can be structured—at will—by locally desorbing hydrogen atoms and thus creating lines [50, 51], spirals [52], or more complex patterns of Si-DBs [53]. However, despite being investigated at room temperature [21], the electronic properties of the various observed structures on the Si(100):H surface such as single silicon dangling bonds or bare silicon dimers are still badly understood although several works have been devoted to their studies [54]. The role of these dehydrogenated features to create, for example, contacts with single molecules is crucial and motivates the investigation of their local electronic properties. The scanning tunneling spectroscopy (STS) is a powerful technique to locally investigate the electronic structure and properties of metallic or semiconductor surfaces. For the clean Si(100) surface, both experimental and theoretical works underline the influence of the temperature as well as the surface reconstruction on the surface electronic structure [55–57]. The type of doping (*n* or *p*) and the nature of dopant atoms (P or As for the *n* type and B for the *p* type) also plays an important role as revealed in the work of Dubois et al. [58] *and more recently in the study of charge state of Si-DB for p-type-doped Si(100):H surfaces* [59]. One of the main differences between the *n* type and the *p* type (i.e., the energy of the Fermi-level location) essentially results in an energy shift of the surface states of the clean Si(100) surface [60]. Although the electronic properties of various dehydrogenated species (single silicon dangling bonds, bare silicon dimers, or lines of silicon dangling bonds) observed on the Si(100):H surface have also been the subject of a number of investigations at room temperature [61, 62], there are only few studies which consider the influence of the type of dopant on these local features [63], especially at low temperature [64]. From a theoretical point of view, the introduction of a dopant atom in the computation cell is specifically justified since studies of this passivated surface at low temperature require using samples with very high dopant concentrations ($\sim 10^{20}$ cm^{-3}). Yet, the electronic structure of the Si(100):H surface and of the ensuing local dehydrogenated Si-DB has rarely been studied at low temperature and for high concentration of dopants. In this section, a low-temperature (5 K) STS study of (i) the clean Si(100) surface, (ii) the hydrogenated Si(100):H surface, (iii) a single silicon dangling bond, (iv) a bare silicon dimer on the Si(100):H surface is performed via dI/dV measurements and compares with numerical simulations. The STS spectra performed at 5 K have the advantage of using low tunnel current (<100 pA) combined with stable measurements that are usually free from thermal broadening and vibronic coupling effects.

These experiments use a low-temperature (5 K) scanning tunneling microscope (LT-STM) under ultrahigh-vacuum (UHV) conditions. The Si(100) samples are As-doped (*n* type), with a resistivity of 5 mΩ cm. After preparing the clean Si

(100)-2×1 surfaces under UHV (base pressure 5×10^{-11} Torr) [65–67], the sample is kept at 650 K and the hydrogenation is performed as previously reported [17, 25]. After hydrogenation, the sample is cooled down and transferred to the STM chamber where all the experiments are performed at 5 K. This procedure allows to obtain a surface with a majority of 2×1 reconstruction areas. The $I(V)$ and $dI(V)/dV$ curves are acquired using an external lock-in amplifier (Stanford Instruments). The measurements are performed with a voltage modulation of 36 mV at a frequency of 863 Hz and with a constant tip height (we repeated the measurements for various tip heights). In order to fully understand the STS spectra and in particular the possible charging of the surface states, we have performed the calculations of the projected density of states (PDOS) of the Si(100):H surface and single silicon dangling bond by taking into account the type of dopant present in the silicon substrate (*n*-type As-doped). The performed calculations have been carried out using the first-principle density-functional theory (DFT) method implemented in the VASP code [68–70]. The geometric optimizations and the electronic properties calculations were performed using the projector-augmented wave approach [71] and the Perdew-Wang (PW91) formulation of the generalized gradient approximations for the exchange and correlation terms [72]. The Si(100) surface is modeled by a periodically repeated (6×6) slab containing four Si layers saturated by hydrogen atoms at the bottom. The vacuum region thickness exceeds 15 Å. For the computation made on the Si(100):H surface, the top surface is recovered by a monohydride phase. In the case of the single dangling bond, one hydrogen atom is removed from the top surface. To simulate the n-doped surface, one substituted As atom is introduced in the third silicon atom layer. Thus, the total number of atoms in the slab is 252 for the fully hydrogenated surface and 251 for the surface with a single dangling bond. The whole system, apart from the H bottom layer, is allowed to relax until the forces on each ion are lower than 0.01 eV/Å. Due to the large size of the employed supercell Brillouin zone, integrations are performed using a single Γ point approximation.

The surface reconstruction of the semiconductors is strongly modified by the bulk electronic properties. In the case of the bare Si(100)-2×1, the surface reconstruction leads to the formation of silicon dimers and the creation of specific surface states related to the presence of σ and σ^*, and π and π^* orbitals [73]. The adsorption of a monolayer of hydrogen atoms on the Si(100) surface induces the breaking of the double Si–Si dimer bond (σ and σ^* orbitals) and the creation of Si–H bonds (π and π^* orbitals), thus keeping the 2×1 reconstruction. Consequently, the STS curves of the clean Si(100) surface and the hydrogenated Si(100):H surface as well as the dehydrogenated surface structures (Fig. 5a and b) might contain specific signatures of these different orbitals. Figure 5d shows the STS curves $I(V)$, $dI(V)/dV$, and $[dI(V)/dV]/(I/V)$, acquired on the bare Si(100) surface (dark blue curves of Fig. 5c, d, and h) and on the hydrogenated Si(100):H surface (black curves of Fig. 5f, g, and e). Note that the STM tip is placed in the middle of the silicon dimer for STS acquisition on the clean Si(100) and the fully hydrogenated surfaces. To identify the various bands that are observed in the curves measured over the Si(100) surface, we compare our curves to the ones that were obtained at

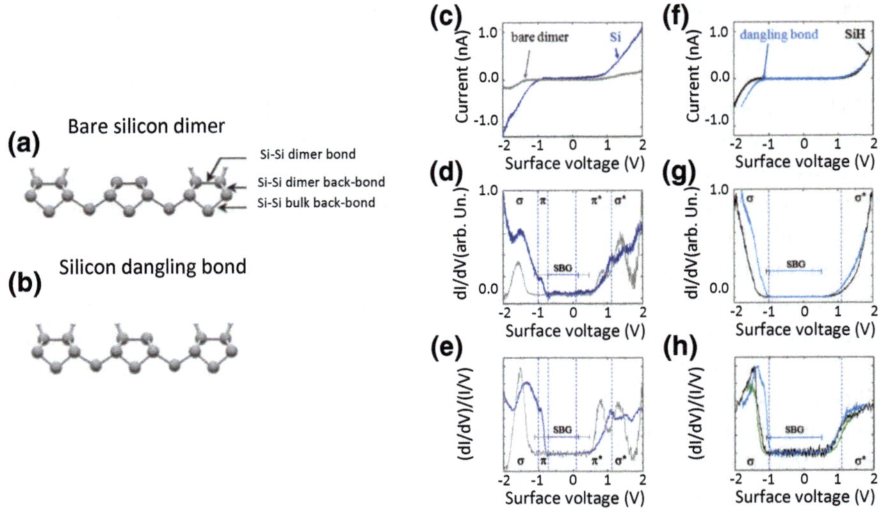

Fig. 5 **a** and **b** Cross-sectional view of the Si(100):H-2×1 surface with a single silicon dangling bond and a bare silicon dimer, respectively. The silicon atoms are in *gray* and the hydrogen atoms in *white*. **c** Comparison of the I(V) curves of the clean Si(100) (*dark blue curve*) and single bare silicon dimer (*gray* curve), and **f** comparison of the I(V) curves of the hydrogenated Si(100):H (*black curve*) and single silicon dangling bond (*light blue curve*). **d** and **g** are the same as dimers and **f**, respectively, but for the d*I*(*V*)/d*V* curves. **e** and **h are** the same as **c** and **f**, respectively, but for normalized (d*I*(*V*)/d*V*)/(*I*/*V*) curves. In **h,** the normalized STS curve recorded on a dihydride Si atom is added (*green curve*) for comparison. Energy zones for the σ, π, π*, and σ* bands are indicated as well as the SBG

room temperature [55, 56] and at 77 K [57, 58]. According to these studies, for a *n*-type substrate, the energy of the Si–Si dimer π band lies between −0.6 and −1.3 eV below the Fermi level, while the energy of the σ band is observed between −1.4 and −2.0 eV. For the unoccupied states, the energy of the Si–Si dimer π* band lies between +0.2 and +1.1 eV above the Fermi level and the energy of the σ* band between +1.4 and +1.9 eV. Note that for these bands, some slight energy shifts can be observed due to the buckling of the Si dimer [i.e., $c(4 \times 2)$ and $p(2 \times 2)$] and to the presence of step edges near the location of the STS curve acquisition. We emphasize that on the clean Si(100) surface, the ensuing σ and σ* bands are located on the Si–Si dimer and on the Si–Si back bonds (BB) of the dimer. Indeed, the Si–Si dimer back bonds formed between the silicon atoms of the first and the second layers play a significant role in the STS curves leading to the observation of an unoccupied band which energy is usually located around +1 eV above the Fermi level for the *n*-type substrate. Considering the above arguments, for the positive voltages, the broadband spreading from +0.2 to +1 eV in the STS curve acquired on the bare Si(100) at 5 K (dark blue curves of Fig. 5d and e) can be assigned to the π* bands of the silicon dimers. Since this latter band is not observed in the STS curve of the Si(100):H surface (Fig. 5g and h), the unoccupied π* character, inherent to the 2×1 reconstruction, is confirmed. In the same way, for negative surface voltages, the signature of the π band can be observed in the STS curve of the bare Si (100) surface (see the band between −0.6 and −1.1 eV in the curves in Fig. 5d and e, while it is missing in the STS curve of the Si(100):H surface (Fig. 5g and h). These attributions for the clean Si(100)-2×1 surface are in agreement with the previous reports in the literature [56, 58]. Contrary to the case of the π and π* bands, there are similar STS bands that can be observed in both Si(100) and Si (100):H spectra. These bands are centered at ~1.4 eV in the occupied states and spread from +1.2 to +2 eV in the unoccupied states. According to the previous work [57, 58], these bands cannot be mainly ascribed to the single Si–Si dimer bond but rather to the Si–Si back bonds of the Si dimer. This assignment is confirmed when STS measurements (green curve in Fig. 5h) are performed on a dihydride silicon atom of the 3×1 reconstruction or of the bow-tie structures. Indeed, on these dehydrogenated structures, the single Si–Si dimer bonds (σ and σ* bands) do not exist anymore. Thus, the fact that the STS spectra acquired on these dihydride structures show very similar STS bands suggests that they are mainly due to σ and σ* orbitals from the Si–Si back bonds instead of the Si–Si dimers. Single silicon dangling bonds (Fig. 6a) and bare silicon dimers (Fig. 6b) can be formed on the Si (100):H-2×1 surface by using the STM tip as a local electron source to desorb H atoms one by one. The corresponding STS curves are shown in Fig. 5. The $dI(V)/dV$ and the normalized spectra (gray curves in Fig. 5d and e) acquired at 5 K on the bare dimer present three distinct narrow bands: one at −1.5 eV in the occupied states and two at +0.8 and +1.3 eV in the unoccupied states. These bands are assigned to the σ, π*, and σ* bands, respectively. The main difference between the spectra of the clean Si(100) surface and the bare silicon dimer on the Si(100):H surface is the absence of the π band in the latter case. In fact, the π band can be seen on the clean Si(100) surface because of a strong delocalization along the silicon dimer rows which enables electron transport across the surface even though the π band has an energy located within the bulk band gap [74, 75]. In the case of the bare silicon dimer on the Si(100):H surface, the π orbital cannot be observed with the STM since the current cannot flow through this localized π orbital. The STS spectra acquired on a single silicon dangling bond formed on the Si(100):H surface are shown in Fig. 5f–h. As compared to the STS spectrum of the hydrogenated Si(100): H surface, the main difference is the energy shift of the σ band toward higher energies for the single silicon dangling bond.

This is most probably due to the upward band bending produced by the negative charging of the silicon dangling bond that we will discuss later. STM topographies of a single silicon dangling bond and a bare silicon dimer created on the Si(100): H-2×1 surface are shown in Fig. 6a and b. In the unoccupied states, the bare dimer is seen as two bright lobes separated by a darker line (Fig. 6b, right panel). Similar features can be observed at room temperature [76] and 120 K [74] with the same apparent symmetry. Such symmetrical feature suggests that the corresponding bare dimer is either not buckled or switching randomly between two stable buckled positions during the STM topography (i.e., the mean position of the dimer is observed). Since our experiments are performed at low temperature (5 K) and with low-perturbing scanning parameters (e.g., +1.7 eV and 110 pA), the second

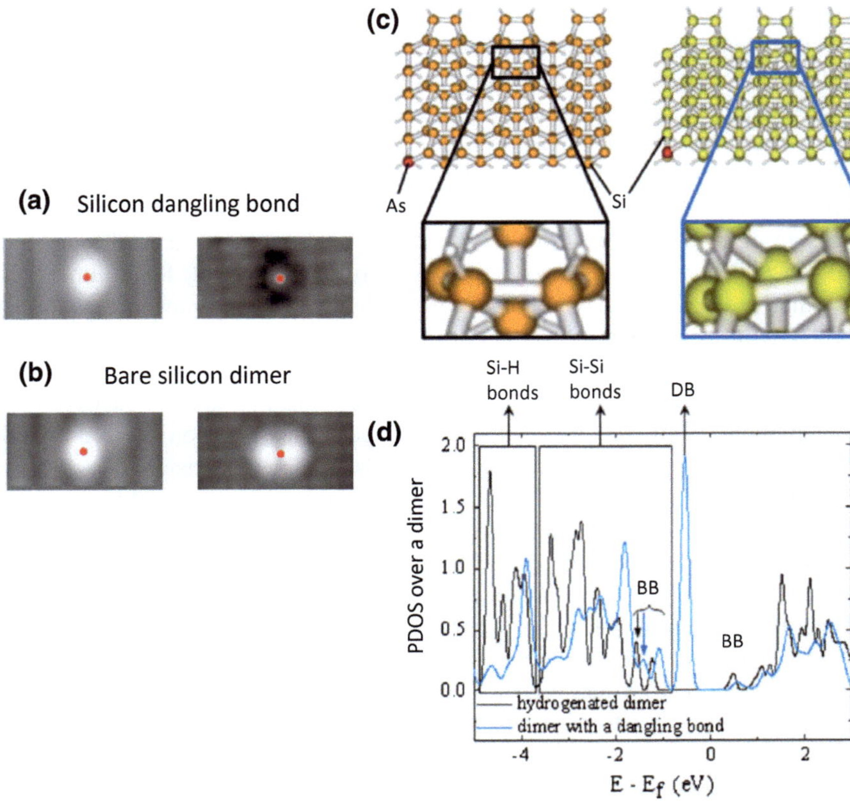

Fig. 6 **a** and **b** STM topographies (3.8 × 2.0 nm^2) on the Si:H surface of a single silicon dangling bond (a, *left*) in the occupied states ($V_S = -1.7$ V and I = 110 pA) and (a, *right*) in the unoccupied states ($V_S = 1.8$ V and I = 110 pA), and of a bare silicon dimer (b, *left*) in the occupied states ($V_S = -1.7$ V and I = 70 pA) and (b, *right*) in the unoccupied states ($V_S = 1.7$ V and I = 70 pA). The *red dots* indicate the location of the tip during the STS curves recording. **c** Calculation slab for the fully hydrogenated surface (*left*) and the same surface with a single silicon dangling bond (*right*). Zooms are realized on the silicon dimer on which are calculated the projected densities of states. **d** Calculated PDOS curves on a hydrogenated silicon dimer (*black* curve), which is composed of two silicon atoms and two hydrogen atoms (*black rectangle* in **c** *left*) and a silicon dimer with a single dangling bond (*light blue* curve), which is composed of two silicon atoms and a sole hydrogen atom (*light blue* rectangle in **c** right). BB corresponds to the Si–Si dimer back bonds

hypothesis can be reasonably ruled out. Acquired at +1.7 eV, the unoccupied state STM topography of the bare dimer feature must correspond to tunnel current flowing through a mixture of π* orbital of the double Si–Si dimer bond and σ* orbital of the Si–Si back bonds of the same dimer. This assignment agrees with the previous work performed at room temperature and 120 K [74]. The bright protrusion observed in the occupied state STM topography of the bare silicon dimer (Fig. 6a left panel) should merely result from tunnel current flowing through the band of the Si–Si dimer back bonds when STM topography is performed at $V_S = -1.7$ V. In the STM topography of Fig. 6b left panel, the single dangling bond appears, for negative surface voltages, as a bright protrusion off center with respect to the dimer row (here on the right side of the dimer row). For positive surface voltages (Fig. 6a, right panel), the STM topography of the single dangling bond shows a smaller gray protrusion surrounded by a dark halo similar to what has been observed at room temperature [63]. This unoccupied STM topography characteristic is often suggested to be due to a charge transfer from the silicon bulk to the dangling bond for n-type-doped samples. Recent investigations point out that the presence of the dark halo might be the result of more complex dynamic processes when electrons flow from the silicon bulk to the DB and the STM tip and inversely [53].

The present investigation also suggests that due to the relatively high energy at which the STM topographies are performed, the interaction between the σ^* orbital of the BB and the dynamically charged state of the DB should be taken into account.

To support our experimental results, we performed the calculation of the density of states on a large silicon slab (1.2 nm^3), allowing us to take into account the presence and the nature of the dopant atom. The insertion of the dopant atom is illustrated in the ball and stick example shown in Fig. 6c. Thus, the relaxation of the slab gives us additional information on the geometry of the silicon dimers. Indeed, for a fully hydrogenated surface (Fig. 6c, left panel), the Si dimers are found to be symmetrical, similar to the previous reports. However, for the case of a silicon dimer with a single silicon dangling bond (Fig. 6c, right panel), the Si dimer is shown to be tilted at an angle of 7.5° with respect to the surface plane which is related to the charge transfer as previously described. In this case, the silicon atom with the charged dangling bond is higher than the Si atom bonded with the H atom. It is interesting to compare this specific dimer geometry with the fully hydrogenated silicon dimer and the individual bare silicon dimer which are both symmetrical. Hence, the fact that the silicon atoms of the bare dimer are at the same height, identical to the fully hydrogenated dimer, suggests that the bare silicon dimer is neutral. This is also supported by the absence of a dark halo surrounding the two bright features observed in Fig. 6b (right panel). For the relaxed structures shown in Fig. 6c, we calculated the PDOS on a single silicon DB which is either completely hydrogenated (black curve in Fig. 6d) or with a single silicon dangling bond (light blue curve in Fig. 6d). In this figure, we can see that the presence of the silicon dangling bond (light blue curve) reveals a specific intense PDOS peak centered at −0.5 eV below the Fermi level. In addition, compared to the fully hydrogenated dimer, there is a clear energy shift, toward higher energies, of the PDOS bands of the Si–Si BB whose energy lies between −0.5 and −1.9 eV below the Fermi level. This is in good agreement with the experimental STS spectra (Fig. 5g and h) on which the surface band gap (SBG) measured over the silicon dangling bond is slightly reduced in the negative voltage part compared to the Si:H surface. As explained before, this energy shift might originate from the upward band bending due to the negatively charged silicon dangling bond. We note that the occupied dangling bond state calculated at −0.5 eV below the Fermi level cannot be seen in the STS spectra of Fig. 5f–h, while a very localized bright feature can be observed on the STM topography of Fig. 6b. This suggests that the energy shift of the back-bond peaks favor the hybridization between the tail of the dangling bond and the BB peaks. Thus, when STM topographies are acquired at a surface voltage $V_S = -1.7$ V (Fig. 6b), the tunnel electrons, which are mainly flowing through the occupied states of the σ orbitals of the Si–Si back bonds, can probe the localized Si-DB feature because of this coupling. The fact that the Si-DB peak centered at −0.5 eV below the Fermi level cannot be detected in the experimental STS spectra is very particular to local electronic transport. Several studies reported that the DOS peak of the Si-DB can be observed at room temperature inside the surface band gap energy range [17]. This difference might suggest that at room temperature, the dangling bond DOS is coupled, through vibronic modes to the substrate atoms, allowing the opening of new electronic transport channels [77]. More recent investigations indicate that the energy of the Si-DB DOS peak is shifted toward the conduction band when probed at positive voltage. This effect induce a specific electronic dynamics in which the Si-DB acts as a quantum dot and is continuously charged and discharged during the tunnel electronic transport from the STM tip to the silicon bulk [78]. In this context, the charge state of the Si-DB for n-type sample is often the subject of controversy since one has to take into account the average charge state that is related to several parameters such as the tunnel current intensity [53], the type of dopant, or the degree of depletion of these dopants within the surface following the sample preparation [27]. These effects will surely be the subject of future investigations.

4 Conclusion

We have presented several methods that are used to passivate the silicon surface with hydrogen. The first method is based on the utilization of wet etching methods with HF or NH_4F acids. In this method, one of the main drawbacks is related to the pollutions of the resulting surface that are introduced in the solvent of the etching solution. Hence, due to its stronger chemical reactivity, this technique is usually less used to prepare the Si(100):H surface rather than the Si(111):H. Additionally, the wet solution is weakly adapted to the Si(100) surface due to the formation of pyramids that forms several surface reconstructions. The second method detailed in this chapter is a UHV compatible method that exploits the reactivity of the 2×1 reconstruction of the Si(100) when exposed to atomic H atoms created either by a gas cracker system or via H_2 ionization with a hot tungsten filament. This process can also be applied on the Si(111) and Ge(100) surfaces. The third method for the passivation of the Si(100) surface is based on a novel discovery that can induce a very local surface passivation process with H_2. It is based on the condensation of molecular hydrogen on the tip apex of a low (5 K) STM that acts as a cold finger. The use of tunnel electrons in the STM junction allows to control the H_2 desorption from the STM tip and its dissociation on the Si(100) surface. The motivations to improve hydrogenation processes via the reactivity of H_2 molecules on semiconductor surface are still nowadays very active and not only on the fundamental side as they can be involved in various applied surface science processes. Indeed, these research topics can easily be related to hydrogen storage at the nanoscale, to quantum information or for the control of chemical reactivity in molecular devices.

References

1. Taylor, H.S.: Chemical reactions at surfaces. Chem. Rev. **9**, 1 (1931)
2. Gourdon, A.: Ch. Joachim, On-Surface Synthesis. Springer, (2014). ISBN: 978-3-319-26598-8
3. Bordo, V.G., Rubhan, H.G.: Optics and Spectroscopy at Surfaces and Interfaces. Wiley (2005). ISBN: 978-3-527-40560-2
4. Kolasinski, K.: Surface Science. Wiley (2008). ISBN: 978-1-1199-9036-9
5. Campbell, I.M., et al.: Catalysis at surfaces. Springer (1988). ISBN: 978-0-412-28970-5
6. Nardi, E., Chen, L., Clair, S., Koudia, M., Giovanelli, L., Feng, X.L., Mulen, K., Abel, M.: On-Surface Reaction between Tetracarbonitrile-Functionalized Molecules and Copper Atoms. J. Of Phys. Chem. C **118**, 27549 (2014)
7. Chiaravalloti F., Dujardin, G., Riedel, D., Pinto, H., Foster, A.: Atomic-scale study of the adsorption of calcium fluoride on Si(100) at low-coverage regime. Phys. Rev. B **84**, 155433 (2001)
8. Lyding, J.W., Hess, K., Kizilyalli, I.C.: Reduction of hot electron degradation in metal oxide semiconductor transistors by deuterium processing. Appl. Phys. Lett. **68**, 2526 (1996)
9. Kolasinski, K.W., Nessler, W., de Meijere, A., Hasselbrink, E.: Hydrogen adsorption on and desorption from Si: Considerations on the applicability of detailed balance. Phys. Rev. Lett. **72**, 1356 (1994)
10. Dürr, M., Biedermann, A., Hu, Z., Höfer, U., Heinz, T.F.: Direct patterning of modified oligonucleotides on metals and insulators by dip-pen nanolithography. Science **296**, 1838 (2002)
11. Labidi, H., Kantorovich, L., Riedel, D.: Atomic-scale control of hydrogen bonding on a bare Si(100)-2×1 surface. Phys. Rev. B **86**, 165441 (2012)
12. Kato, H., Taoka, T., Nishikata, S., Sazaki, G., et al.: Preparation of an Ultraclean and Atomically Controlled Hydrogen-Terminated Si(111)-(1× 1) Surface Revealed by High Resolution Electron Energy Loss Spectroscopy, Atomic Force Microscopy, and Scanning Tunneling Microscopy: aqueous NH4F Etching Process of Si(111). Jap. J. Appl. Phys. **46**, 5701 (2007)
13. Laracuente, A.R., Baker, L.A., Whitman, L.J.: UHV characterization of ambient-dosed hydrogen-terminated Si(001). Surf. Sci. **602**, 3 (2008)
14. This process needs to be performed with ultra-high purity materials to avoid recontamination of the bare silicon surface
15. Arapkina, L.V., Krylova, L.A., Chizh, K.V., Chapnin, V.A., Uvarov, O.V., Yuryev, V.A.: Application of hydrogenation to low temperature cleaning of the Si(001) surface in the processes of molecular-beam epitaxy: investigation by scanning tunneling microscopy, reflected high-energy electron diffraction, and high resolution transmission electron microscopy. J. of Appl. Phys. **112**, 014311 (2012)
16. Storozhevykh, M.S., Arapkina, L.V., Yuryev, V.A.: Evidence for Kinetic Limitations as a Controlling Factor of Ge Pyramid Formation: a Study of Structural Features of Ge/Si(001) Wetting Layer Formed by Ge Deposition at Room Temperature Followed by Annealing at 600 °C. Nano. Res. Lett. **10**, 295 (2015)

17. Boland, J.J.: Scanning tunnelling microscopy of the interaction of hydrogen with silicon surfaces. Adv. Phys. **42**, 129 (1993)
18. Riedel, D., Delattre, R., Borisov, A., Teperik, T.: A Scanning Tunneling Microscope as a Tunable Nanoantenna for Atomic Scale Control of Optical-Field Enhancement. Nano Lett. **10**, 3857 (2010)
19. Bellec, A., Ample, F., Riedel, D., Dujardin, G., Joachim, Ch.: Imaging Molecular Orbitals by Scanning Tunneling Microscopy on a Passivated Semiconductor. Nano Lett. **9**, 144 (2009)
20. Christmann, K.: Interaction of hydrogen with solid surfaces. Surf. Sci. Rep. **9**, 1 (1988)
21. Mayne, A.J., Riedel, D., Comtet, G., Dujardin, G.: Atomic-scale studies of hydrogenated semiconductor surfaces. Prog. Surf. Sci. **81**, 1 (2006)
22. Oura, K., Lifshits, V.G., Saranina, A.A., Zotova, A.V., Katayamaa, M.: Some aspects of the surface behaviour of silicon. Surf. Sci. **35**, 1 (1999)
23. Bischler, U., Bertel, E.: Simple source of atomic hydrogen for ultrahigh vacuum applications. J. Vac. Sci. Technol. A **11**, 458 (1993)
24. This is related to the bright red color of the silicon during the short heating period at 1100 °C
25. Bellec, A., Riedel, D., Dujardin, G., Rampotis, N., Kantorovich, L.: Dihydride dimer structures on the Si(100):H surface studied by low-temperature scanning tunneling microscopy. Phys. Rev. B **78**, 165302 (2008)
26. Suwa, Y., Matsuura, S., Fujimori, M., Heike, S., Onogi, T., Kajiyama, H., Hitosugi, T., Kitazawa, K., Uda, T., Hashizume, T.: Dopant-Pair Structures Segregated on a Hydrogen-Terminated Si(100) Surface. Phys. Rev. Lett. **90**, 156101 (2003)
27. Labidi, Hatem, Taucer, Marco, Rashidi, Mohammad, Koleini, Mohammad, Livadaru, Lucian, Pitters, Jason, Cloutier, Martin, Salomons, Mark, Wolkow, Robert A.: Scanning tunneling spectroscopy reveals a silicon dangling bond charge state transition. New J. Phys. **17**, 073023 (2015)
28. Riedel, D., Bocquet, M.L., Lesnard, H., et al.: Selective Scanning Tunnelling Microscope Electron-Induced Reactions of Single Biphenyl Molecules on a Si(100) Surface. Journ. Of Am. Chem. Soc. **131**, 7344 (2009)
29. Riedel, D.: Single molecule manipulation at low temperature and laser scanning tunneling photo-induced processes analysis through time-resolved studies. J. Phys. Cond. Mat. **22**, 264009 (2010)
30. Dürr, M., Höfer, U.: Dissociative adsorption of molecular hydrogen on silicon surfaces. Surf. Sci. Rep. **61**, 465 (2006)
31. Bellec, A., Riedel, D., Dujardin, G., Boudrioua, O., Chaput, L., Stauffer, L., Sonnet, Ph: Electronic properties of the n-doped hydrogenated silicon (100) surface and dehydrogenated structures at 5 K. Phys. Rev. B **80**, 245434 (2009)
32. Hirschfelder, J.O., Curtiss, C.F., Bird, R.B.: Molecular Theory of Gases and Liquids. Wiley, New York (1967)
33. Chuikov, B.A., Dvurechenskikh, V.V., Osovskii, V.D., Ptushinskii, Y.G., Sukretnyi, V.G.: Adsorption of H2 and D2 on W(111) and W(110) surfaces at low temperatures (down to 5 K). Surf. Sci. **285**, 75 (1993)
34. Bratu, P., Brening, W., Groβ, A., Hartman, M., Höefer, U., Kratzer, P., Russ, R.: Reaction dynamics of molecular hydrogen on silicon surfaces. Phys. Rev. B **54**, 5978 (1996)
35. Watamori, M., Naitoh, M., Morioka, H., Maeda, Y., Oura, K.: Low temperature adsorption of hydrogen on Si(111) and (100) surfaces studied by elastic recoil detection analysis. Appl. Surf. Sci. **82**, 417 (1994)
36. Kolasinski, K.: Dynamics of hydrogen interactions with Si(100) and Si(100) surfaces. Intern. J. Mod. Phys. B **9**, 2753 (1995)
37. Henriksson, K.O.E., Vörtler, K., Dreißigacker, S., Nordlund, K., Keinonen, J.: Sticking of atomic hydrogen on the tungsten (0 0 1) surface. Surf. Sci. **600**, 3167 (2006)
38. Chuikov, B.A., Osovskii, V.D., Ptushinkii, Y.G., Sukretnyi, V.G.: Influence of the flux of H2 and D2 molecules on kinetics of low-temperature (down to 5 K) adsorption on the W(110). Surf. Sci. **473**, 143 (2001)

39. Ptushinskii, Y.G.: Low-temperature adsorption of gases on metal surface. Low temp. Phys. **30**, 1063 (2004)
40. Norlander, P., Holmberg, C., Harris, J.: Physisorption interaction of H2 with noble metals. Surf. Sci. **152**, 702 (1985)
41. Henriksson, K.E.O., et al.: Sticking of atomic hydrogen on the tungsten (0 0 1) surface. Surf. Sci. **600**, 3167 (2006)
42. Nahm, T.U., Gomer, R.: Desorption of H2 from W(110) and Fe covered W(110) surfaces. Surf. Sci. **380**, 434 (1997)
43. Andersson, S., Svensson, K.: Electron-Induced Desorption of Physisorbed H2 via Resonance Vibrational Excitation. Phys. Rev. Lett. **102**, 156104 (2010)
44. Sharp, T.E.: Potential-energy curves for molecular hydrogen and its ions. Atomic data **2**, 119 (1971)
45. Lang, N.D., Kohn, W.: Theory of Metal Surfaces: Induced Surface Charge and Image Potential. Phys. Rev B **7**, 3541 (1973)
46. Golser, R., Gnaser, H., Kutshera, W., Priller, A., Steier, P., Wallner, A., Čížzek, M., Horáček, J., Domcke, W.: Experimental and Theoretical Evidence for Long-Lived Molecular Hydrogen Anions H_2^- and D_2^-. Phys. Rev. Lett. **94**, 223003 (2005)
47. Yengui, M., Pinto, H., Leszczynski, J., Riedel, D.: Atomic scale study of corrugating and anticorrugating states on the bare Si(100) surface. J. Phys. Cond. Matt. **27**, 045001 (2015)
48. Manzano, C., Soe, W.-H., Kawai, H., Saeys, M., Joachim, C.: Origin of the apparent (2×1) topography of the Si(100)−c(4×2) surface observed in low-temperature STM images. Phys. Rev. B **83**, 201302 (2011)
49. Ample, F., Joachim, C.: The chemisorption of polyaromatic hydrocarbons on Si(100):H dangling bonds. Surf. Sci. **602**, 1563 (2008)
50. Doumergue, P., Pizzagalli, L., Joachim, C., Altibelli, A., Baratoff, A.: Conductance of a finite missing hydrogen atomic line on Si(001)-(2×1)-H. Phys. Rev. B **59**, 15910 (1999)
51. Hitosugi, T., Heike, S., Onogi, T., Hashizume, T., Watanabe, S., Li, Z.Q., Ohno, K., Kawazoe, Y., Hasegawa, T., Kitazawa, K.: Jahn-Teller Distortion in Dangling-Bond Linear Chains Fabricated on a Hydrogen-Terminated Si(100)- 2×1 Surface. Phys. Rev. Lett. **82**, 4034 (1999)
52. Hitosugi, T., Hashizume, T., Heike, S., Kajiyama, H., Wada, Y., Watanabe, S., Hasegawa, T., Kitazawa, K.: Scanning tunneling microscopy/spectroscopy of dangling-bond wires fabricated on the Si (100)-2× 1-H surface. Appl. Surf. Sci. **130**, 340 (1998)
53. Schofield, S.R., Studer, P., Hirjibehedin, C.F., Curson, N.J., Aeppli, G., Bowler, D.R.: Quantum engineering at the silicon surface using dangling bonds. Nat. Com. **4**, 1649 (2013)
54. Livadaru, L., Pitters, J., Taucer, M., Wolkow, R.A.: Theory of nonequilibrium single-electron dynamics in STM imaging of dangling bonds on a hydrogenated silicon surface. Phys. Rev. B **84**, 205416 (2011)
55. Hata, K., Shibata, Y., Shigekawa, H.: Fine electronic structure of the buckled dimers of Si(100) elucidated by atomically resolved scanning tunneling spectroscopy and bias-dependent imaging. Phys. Rev. B **64**, 235310 (2001)
56. Okada, H., Fujimoto, Y., Endo, K., Hirose, K., Mori, Y.: Detailed analysis of scanning tunneling microscopy images of the Si(001) reconstructed surface with buckled dimers. Phys. Rev. B **63**, 195324 (2001)
57. Perdigão, L., Deresmes, D., Grandidier, B., Dubois, M., Delerue, C., Allan, G., Stiévenard, D.: Semiconducting Surface Reconstructions of p-Type Si(100) Substrates at 5 K. Phys. Rev. Lett. **92**, 216101 (2004)
58. Dubois, M., Perdigao, L., Delerue, C., Allan, G., Grandidier, B., Deresmes, D., Stiévenard, D.: Scanning tunneling microscopy and spectroscopy of reconstructed Si(100) surfaces. Phys. Rev. B **71**, 165322 (2005)
59. Bellec, A., Chaput, L., Dujardin, G., Riedel, D., Stauffer, L., Sonnet, Ph: Reversible charge storage in a single silicon atom. Phys. Rev. B **88**, 241406 (2013)
60. Luth, H.: Surfaces and interfaces of solid materials. Springer, Berlin (1995)

61. Bobrov, K., Comtet, G., Dujardin, G., Hellner, L.: Electronic structure of partially hydrogenated Si(100)-(2×1) surfaces prepared by thermal and nonthermal desorption. Phys. Rev. Lett. **86**, 2633 (2001)
62. Haider, M.B., Pitters, J.L., DiLabio, G.A., Livadaru, L., Mutus, J.Y., Wolkow, R.A.: Controlled coupling and occupation of silicon atomic quantum dots at room temperature. Phys. Rev. Lett. **102**, 046805 (2009)
63. Liu, L., Yu, J., Lyding, J.W.: Nanoptterning–From ultralarge- scale integration to Biotechnology. In: MRS Symposia Proceedings No. 705, p. Y6.6.1. Materials Research Society, Pittsburgh (2002)
64. Riedel, D., Lastapis, M., Martin, M.G., Dujardin, G.: Influence of tip-surface interactions and surface defects on Si(100) surface structures by low-temperature (5 K) scanning tunneling microscopy. Phys. Rev. B **69**, 121301(R) (2004)
65. Chiaravalloti, F., Dujardin, G., Riedel, D.: Atomic scale control of hexaphenyl molecules manipulation along functionalized ultra-thin insulating layer on the Si(1 0 0) surface at low temperature (9 K). J. Phys. Cond. Matt. **27**, 054006 (2015)
66. Riedel, D., Mayne, A.J., Dujardin, G.: Atomic-scale analysis of hydrogen bond breaking from Si(100):H induced by optical electronic excitation. Phys. Rev. B **72**, 233304 (2005)
67. Bellec, A., Riedel, D., Dujardin, G.: Dihydride dimer structures on the Si(100):H surface studied by low-temperature scanning tunneling microscopy. Phys. Rev. B **78**, 165302 (2008)
68. Kresse, G., Hafner, J.: Ab initio molecular-dynamics simulation of the liquid-metal–amorphous-semiconductor transition in germanium. Phys. Rev. B **49**, 14251 (1994)
69. Kresse, G., Furthmuller, J.: Efficiency of ab-initio total energy calculations for metals and semiconductors using a plane-wave basis set. Comput. Mater. Sci. **6**, 15 (1996)
70. Kresse, G., Furthmuller, J.: Efficient iterative schemes for ab initio total-energy calculations using a plane-wave basis set. Phys. Rev. B **54**, 11169 (1996)
71. Blochl, P.E.: Projector augmented-wave method. Phys. Rev. B **50**, 17953 (1994)
72. Wang, Y., Perdew, J.P.: Correlation hole of the spin-polarized electron gas, with exact small-wave-vector and high-density scaling. Phys. Rev. B **44**, 13298 (1991)
73. Kubby, J.A., Boland, J.J.: Scanning tunneling microscopy of semiconductor surfaces. Surf. Sci. Rep. **26**, 61 (1996)
74. Chen, D., Boland, J.J.: Chemisorption-induced disruption of surface electronic structure: Hydrogen adsorption on the Si(100)−2×1 surface. Phys. Rev. B **65**, 165336 (2002)
75. Sagisaka, K., Fujita, D.: Standing waves on Si(100) and Ge(100) surfaces observed by scanning tunneling microscopy. Phys. Rev. B **72**, 235327 (2005)
76. Boland, J.J.: Evidence of pairing and its role in the recombinative desorption of hydrogen from the Si(100)-2×1 surface. Phys. Rev. Lett. **67**, 1539 (1991)
77. Berthe, M., Urbieta, A., Perdigão, L., Grandidier, B., Deresmes, D., Delerue, C., Stiévenard, D., Rurali, R., Lorente, N., Magaud, L., Ordejón, P.: Electron transport via local polarons at interface atoms. Phys. Rev. Lett. **97**, 206801 (2006)
78. Livadaru, L., Pitters, J., Taucer, M., Wolkow, R.A.: Theory of nonequilibrium single-electron dynamics in STM imaging of dangling bonds on a hydrogenated silicon surface. Phys. Rev. B **84**, 205416 (2011)

Nanopackaging of Si(100)H Wafer for Atomic-Scale Investigations

Delphine Sordes, Aurélie Thuaire, Patrick Reynaud, Caroline Rauer, Jean-Michel Hartmann, Hubert Moriceau, Emmanuel Rolland, Marek Kolmer, Marek Szymonski, Corentin Durand, Christian Joachim, Séverine Chéramy and Xavier Baillin

Abstract Ultra-high vacuum (UHV) investigations have demonstrated a successful development of atomic nanostructures. The scanning tunneling microscope (STM) provides surface study at the atomic scale. However, the surface preparation is a crucial experimental step and requires a complex protocol conducted in situ in a UHV chamber. Surface contamination, atomic roughness, and defect density must be controlled in order to ensure the reliability of advanced UHV experiments. Consequently, a packaging for nanoscale devices has been developed in a microelectronic clean room environment enabling the particle density and contaminant concentration control. This nanopackaging solution is proposed in order to obtain a Si(001)-(2×1):H reconstructed surface. This surface is protected by a temporary silicon cap. The nanopackaging process consists in a direct bonding of two passivated silicon surfaces and is followed by a wafer dicing step into 1-cm^2 dies. Samples can be stored, shipped, and in situ opened without any additional treatment. A specific procedure has been developed in order to open the nanopackaged samples in a UHV debonder, mounted in the load-lock chamber of a

D. Sordes (✉)
CEATech Midi-Pyrénées, Campus INSA, Bât.GM 135 Avenue de Rangueil, 31400 Toulouse, France
e-mail: delphine.sordes@cemes.fr

A. Thuaire · P. Reynaud · C. Rauer · J.-M. Hartmann · H. Moriceau · E. Rolland · S. Chéramy · X. Baillin
University Grenoble Alpes, 38000 Grenoble, France

A. Thuaire · P. Reynaud · C. Rauer · J.-M. Hartmann · H. Moriceau · E. Rolland · S. Chéramy · X. Baillin
CEA-LETI, MINATEC Campus, 38054 Grenoble, France

M. Kolmer · M. Szymonski
Faculty of Physics, Astronomy and Applied Computer Science, Centre for Nanometer-Scale Science and Advanced Materials, NANOSAM, Jagiellonian University, St. Lojasiewicza Street, no. 11, 30-348 Krakow, Poland

C. Durand · C. Joachim
PicoLab, CEMES/CNRS, 29 Rue Marvig, BP 94347, 31055 Toulouse Cedex, France

M. Kolmer and C. Joachim (eds.), *On-Surface Atomic Wires and Logic Gates*, Advances in Atom and Single Molecule Machines,
DOI 10.1007/978-3-319-51847-3_2

low-temperature STM system (LT-STM). Statistical large scan LT-UHV-SEM images and LT-UHV-STM images have been obtained enabling the surface study at the atomic resolution.

1 Introduction

Nowadays, science cannot progress without collaboration between different partners. The particular skill provided by each research laborator or industrial member of a consortium is the key success criteria for a collaborative project. One of the limits for these collaborations is the possibility to exchange samples without degradation due to transportation. Typically, surface contamination can happen during the surface shipment between different laboratories. Indeed, impurities can be trapped by the sample due to the atmospheric air exchange such as hydrocarbons or particles with the sample surface. Consequently, a protection of a functionalized surface or an object on a surface must be carried out in order to conserve its integrity. The packaging for nanoscale device, named "nanopackaging" [1] aims at encapsulating organic or inorganic material at the nanoscale in an appropriate environment and connecting these objects. The nanopackaging field is particularly relevant for atomic-scale devices fabrication and surface atomic wires studies which the specifications concerning the surface properties must deal with the lowest roughness and smallest defect density as possible. These surface structure studies are mostly conducted using the scanning tunneling microscopy (STM) used also, for example, to measure the electronic conductance of a single molecule [2, 3] or to study the surface conductance of atomic-scale systems [4]. However, a well-atomically ordered surface, passivated with light atoms like hydrogen having a low defect density and large terraces, is now requested.

Consequently, the nanopackaging solution has been proposed and one solution is based on a silicon chip sliced from a 200-mm silicon wafer coming from the development nanotechnology clean room [5, 6]. These developments aim at both protecting and connecting an object at the nanoscale (nano-object), an intramolecular circuit, or an atomic-scale circuit.

The process must change the silicon surface in order to obtain a reconstructed Si(001)-(2×1):H surface. For this purpose, two steps need to be achieved: first, the surface large terrace preparation performed in a clean room, and second the surface modification conducted in a dedicated environment permitting the reduction of defect particles and volatiles contaminants.

The surface preparation aims at the hydrogen saturation of the silicon surface followed by a thermal treatment in order to rebuild the silicon surface [7]. Then, the nanopackaging process is performed by a direct wafer bonding of two passivated silicon surfaces [8], permitting a hermetic seal, and consequently, a surface preservation against the atmosphere. Finally, the surface-to-the-nano-object connection has been ensured by a backside connection from the nanoscale device to the macroscale characterizing tool.

STM studies are generally performed in laboratories. In this way, a 200-mm wafer dicing step has been included enabling to obtain 1-cm^2 dies. The protected surface is a determining parameter for STM applications because the temporary cap allows keeping a stable Si(001):H surface. These samples can be stored during few months out of the clean room and be opened only when the manipulation is planned. No contamination is normally coming into play during the transportation between the clean room and the laboratories. That is why, these samples are especially used for collaborative project where an exchange between different laboratories takes place. The Si(001)-(2$\times$1):H surface is particularly appreciated for ultra-high vacuum (UHV) applications because of the well-ordered SiH atoms enabling the surface atomic wire construction using the STM atom-by-atom manipulation mode [9].

These samples are debonded in a UHV opening holder, specially designed and mounted in a load-lock chamber attached to a cryogenic low-temperature STM system (LT-STM). A crucial factor in the debonding procedure is the way of mounting the sample on the holder. Therefore, a method has been developed and a reproducible opening protocol has been established. Indeed, the samples are transferred immediately after the debonding process to the LT-STM sample stage, where they are cooled down to 4 K. The reconstructed Si(001)-(2$\times$1):H surface can be characterized, and LT-UHV-STM investigations are thus able to be conducted at the atomic scale. This chapter reports a detailed study at the atomic resolution of such surface thanks to statistical large scan obtained with the last generation of LT-UHV-STM equipment.

2 Nanopackaged Chip Fabrication

2.1 Packaging Solution for Nanoscale Objects

The packaging solution dedicated to protect and connect an object or a circuit at the nanoscale consists in a structure based on silicon microelectronics technologies. A platform has been developed on a silicon-on-insulator (SOI) substrate including a reconstructed Si(001):H(2$\times$1) surface, dedicated to fabricate an atomic-scale device [9, 10], as depicted in Fig. 1.

This well-ordered and hydrogenated silicon surface is particularly suitable for molecule grafting or surface nanoscale device fabrication. The nano-object is protected by a silicon temporary cap providing a hermetic encapsulation thanks to a direct bonding between two silicon wafers. This structure has been developed in 200 mm wafer diameter but atomic-scale investigations are mostly conducted in samples of a few centimeters; consequently, the bonded wafers have been diced in order to obtain 10 $\times$ 10 mm^2 or 10 $\times$ 12 mm^2 or 10 $\times$ 14.2 mm^2 chips size as requested for the STM characterization. Previously, a bevel (cf. Sect. 2.2.2) has been etched at the edge of the silicon cap in order to facilitate the cap removal.

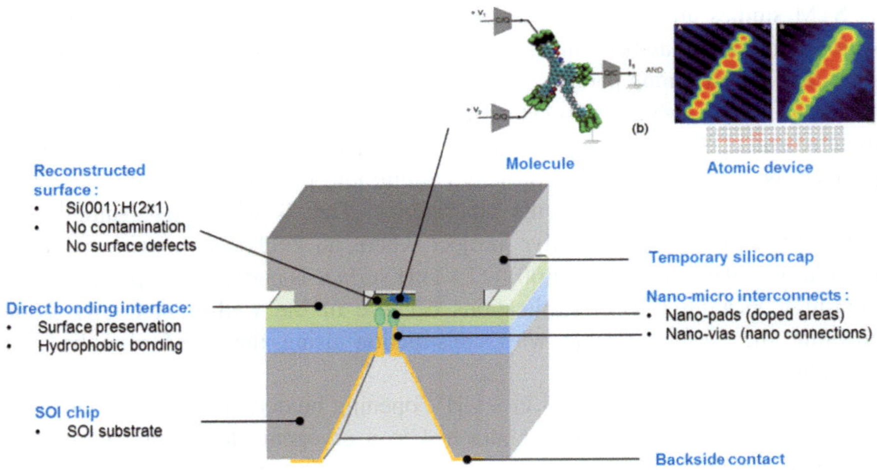

Fig. 1 The 1 × 1 cm^2 chip allows to connecting a molecule or an atomic-scale device on the functionalized Si(001):H(2×1) surface. A silicon temporary cap containing a cavity provides the hermetic seal and ensures the nano-object protection. The connection from the object at the nanoscale is ensured by nano–microinterconnects toward the backside of the chip on which metal pads provide a connection to an electrical external source

As described in Fig. 1, thanks to this new kind of characterizing platform, the electrical properties of the nano-object can be studied, particularly because of the electrical connection with the backside of the chip. Indeed, the nano-object is located between the two implanted regions in the silicon layer and connected to nanovias through the box (silicon dioxide) layer of the silicon-on-insulator (SOI) substrate. The connection at the microscale is ensured by metal lines along a cavity achieved in the backside of the SOI substrate. The metal lines are extended until the backside of the chip on which metal pads are deposited providing the contact by submicronic probes for electrical studies of the nano-object.

However, this chapter is focused on the critical process module development consisting in the silicon surface improvement of the Si(001) surface damaged during the different steps of the process fabrication. A particular attention has been paid to the preparation and the packaging of this surface and especially concerning the roughness, the topology, or the particles density. These parameters must be controlled in order to preserve the surface at the atomic resolution required for a STM imaging.

2.2 *Nanopackaged Chip Fabrication*

The nanopackaging process is composed of elementary modules developed on a SOI substrate composed of a 400-nm-thick SiO_2 layer and 205-nm-thick monocrystalline silicon layers. The silicon material properties have an important impact on the surface quality after surface reconstruction. In a perfectly cut wafer, an atomically flat surface will be obtained. At the opposite, wafer atomic terraces are formed when the flatness is not exactly along [110] and kinks can then be created. The misorientation angle which is less than few degrees from the (100) crystal plane is an important parameter for direct wafer bonding control [11], epitaxy models, and atomic-scale investigations. All the presented work here has been lead using a silicon (001) surface with a miscut angle ranging from 0.01° to 0.4°. Concerning the silicon doping, another important parameter not detailed more in here, the study has been conducted using P-type silicon with a resistivity of 8–10 Ω.cm.

The whole nanopackaging process flow can be divided in four mean technological modules as represented in Fig. 2:

Step 1–2: the first step consists in a chemical cleaning in order to remove particles and contaminants before the surface reconstruction treatment. This step is followed by surface reconstruction in a reduced pressure chemical vapor deposition reactor (RP-CVD) using hydrogen gas. It yields (2 × 1) reconstructed, H-terminated silicon (001) surfaces.

Steps 3–4: the protection of the last passivated reconstructed silicon surface is provided by hydrophobic direct bonding with a second reconstructed surface wafer with the same quality (temporary silicon cap). The bonded wafers are diced into 1-cm^2 samples in order to be loaded in a STM holder.

Steps 5–6: the surface of the samples is characterized and the nanopackaged chip is shipped to laboratories where the temporary cap is removed in a UHV chamber.

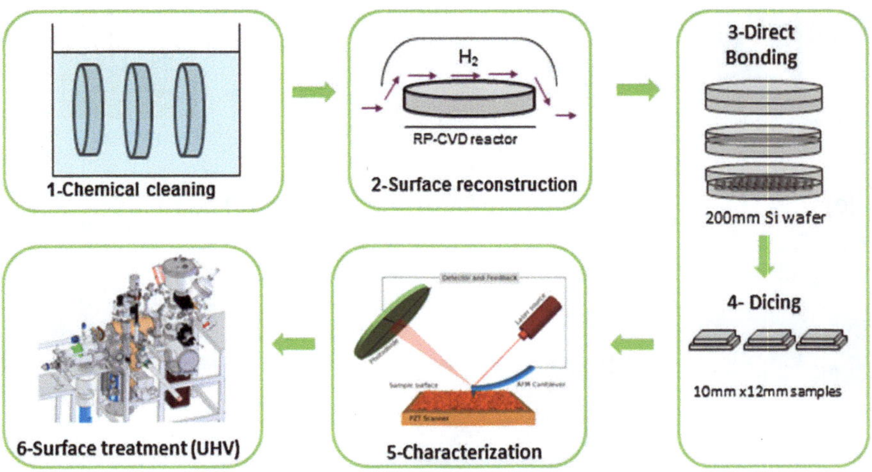

Fig. 2 Process flow description of nanopackaged SiH samples composed of a chemical cleaning, a surface reconstruction, a bonding of a silicon cap for surface preservation, and a dicing in 1-cm^2 pieces. The samples have been characterized before the shipment and a surface investigation in a UHV chamber can be started after the temporary cap removal

LT-UHV-STM investigations can thus be conducted at the atomic scale, especially in order to create a surface atomic wire.

When the electrical measurement on a nano-object is to be achieved, an optional step can be implemented consisting in its connection by the backside of a 200-mm substrate in which cavities have been etched. Two doped regions enabling the electrical continuity of the nano-object have been conducted in the SOI layer. These areas (nanopads) are connected with vias through the oxide layer of the SOI substrate at the nanoscale. These nanovias have been fabricated using a focused ion beam (FIB) which aims at etching the vias and filling using a tungsten plasma gun [12]. Finally, the interconnections between the nanovias and the backside metal pads at the submicronic scale have been obtained with tungsten lines deposited along the edge side of the cavity as illustrated at Fig. 3.

The interconnection development has been optimized with the fabrication of different matrices of 3 × 3 nanovias up to 7 × 7 nanovias with vias diameter of 100 nm, and a depth of 400 nm demonstrated that a high density of interconnections (4.106 vias/mm^2) has been successfully achieved [5]. Firstly, electrical measurements performed on a nanovias matrix have demonstrated that nanovias were clearly insulated from each other. Secondly, concerning the nano-object connection itself, it has been demonstrated that the distance between two implanted areas can be decreased down to 30 nm. This last distance is in the range accommodating the actual small extension of intramolecular circuit or of surface atomic-scale circuits [6]. The nano-object positioning between the doped areas is complex; consequently, in order to localize the implanted regions at the nanoscale, specific marks have been designed in the front side of the wafer.

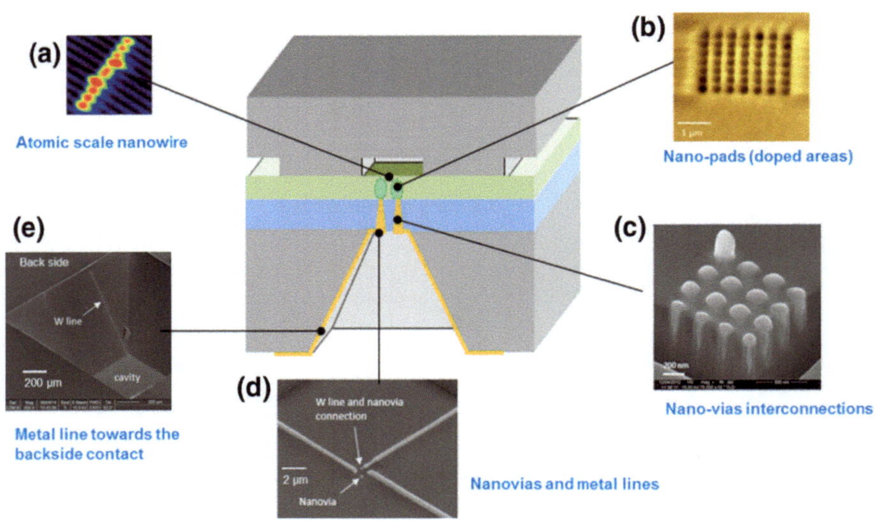

Fig. 3 Interconnections of an atomic-scale dangling bonds nanowire **a** composed of: **b** doped areas (nanopads) linked to **c** nanovias trough a 400-nm-thick oxide layer, connected to **d** tungsten metal lines extended in the side of the cavities **e** toward the backside of the wafer

The backside interconnection step of this nanopackaging process is not always achieved, and in this case, only steps 1–4 will be achieved. For example, the presented STM analysis of the Si(001)-(2×1):H in this chapter does not need backside connections.

2.2.1 Reconstruction of the Silicon Surface

The first step of the process flow consists in Si(001)-(2×1):H front side surface preparation. Surface preparation is indeed of prime importance in order to obtain a Si(001) surface fit for hydrophobic bonding, atomic-scale engineering, or epitaxy. In the latter case, the goal is indeed to duplicate the atomic rows of the substrate into the layer. The presence on the surface of amorphous silicon oxide layers, polymers, or etching residues has then a crippling effect on the epitaxy quality. Depending on the nature of the substrate, two strategies may be used to obtain contamination-free Si surfaces fit for use.

The starting substrate may be bulk Si or a SOI substrate. It is then possible to carry out high-temperature bakes, during which dozens of liters of ultra-pure hydrogen, typically only a few parts per billion of impurities thanks to dedicated gas purifiers, are injected at reduced pressure into the epitaxy chamber in order to remove the 0.8–1-nm-thick layer of native or chemical SiO_2 which is initially present on the surface and smooth it. The simplified chemical reaction for the removal of the oxide layer is as follows:

$$SiO_2(s) + 2H_2(g) \rightarrow Si(s) + 2H_2O(g)$$

The need for high thermal budgets (typically higher than 1050 °C for 1 min, 1100 °C for 2 min as the CEA-LETI standard) prohibits the use of this surface preparation on patterned wafers (i.e., with active silicon zones surrounded by dielectrics) or ion implanted substrates. The active zones could facet due to strain generated by masking dielectrics, especially in a shallow trench isolation (or STI) configuration. An exo-diffusion of preimplanted atoms and a consequent autodoping of the epitaxial layer may also take place [13]. This type of bake is otherwise prohibited for thin Si films (e.g., less than 20 nm) on top of the buried oxides. If these criteria are not respected, moat recess and islanding of the films will take place [14, 15].

The surface preparation described above (where a native or chemical oxide present on the surface is removed by an in situ H_2 bake in suitable conditions) is thus far from being universal given the high thermal budgets needed. In order to minimize the latter, the following sequence is then typically used:

(i) Native SiO_2 is removed by a "HF-last" wet cleaning [16]. During the next-to-last step of such a cleaning, the wafer is dipped in hydrofluoric acid diluted in deionized and deoxygenated water (dilution typically between 0.2

and 1%), in order to etch the SiO_2 layer. The dissolution reaction is as follows [17]:

$$SiO_2 + 4HF \rightarrow 2H^+ + SiF_6^{2-} + 2H_2O$$

The etching rate of thermal SiO_2, which is the most resistant of all, is around 1.2–1.4 nm.min^{-1} for 0.2% HF. During the last step, the wafer is rinsed in deionized and deoxygenated water to remove all the traces of HF. The wafer is then dried by for instance isopropyl alcohol vapors, before being loaded as quickly as possible in the load-lock chambers of the epitaxy tool, in an inert atmosphere (purified N_2 gas). After such a wet cleaning, approximately 85% of the Si dangling bonds are occupied by hydrogen atoms. The remaining 15% are mainly occupied by fluorine atoms as well as oxygen and carbon contaminants [18, 19].

(ii) An H_2 in situ bake then takes place at temperature above 800 °C to remove all O, F, or C surface contaminants [16]. A (2 × 1) reconstructed surface perfectly passivated by hydrogen atoms is then obtained which is ideal for epitaxy, hydrophobic bonding or atomic-scale engineering.

A "HF-last" surface is only stable for a few dozens of minutes up to a couple of hours before the regrowth of a native oxide a few angstroms thick can be detected by spectroscopic ellipsometry. Minimizing the time between a "HF-last" wet cleaning and the loading of wafers inside chambers purged with high purity inert gases (such as N_2) is thus mandatory [20]. A perfectly hydrogen-passivated Si surface obtained by a high-temperature H_2 bake is by contrast stable for several days [21].

Concerning the reconstructed surface preparation, we have thus proceeded as follows: after a "HF-last" wet cleaning, we have proceeded with 20 Torr H_2 bakes in order to get rid of residual contaminants and formed (2 × 1)—reconstructed Si surfaces as atomically smooth as possible. Temperatures and durations varied depending on the nature of the substrate: 1100 °C and 2 min for bulk Si(001), to be compared with 950 °C and 2 to 7 min for SOI substrates (for the reasons mentioned above).

In the former case [e.g., on bulk Si(001)], we have deposited at 950 °C, 20 Torr a few tens of nm of Si on top of the annealed surfaces. We have used to that end the lowest SiH_2Cl_2 mass-flow deliverable in a 200-mm Applied Materials Epi Centura 5200 epitaxy tool. The goal was to remove the Si substrate's starting surface (which might be damaged at the atomic scale), at high temperature and with a low growth rate (75 nm min^{-1} for $F(SiH_2Cl_2)/F(H_2) = 0.00125$, typically). We thus hoped to promote the surface mobility of Si adatoms and have some step flow growth. The "bulk" process ended up with a 5-min, H_2 bake at 950 °C, 20 Torr in order to obtain a stabilized (2 × 1)H—reconstructed Si surface as perfect at the atomic scale as possible.

An atomic force microscopy (AFM) study has been performed on Si(100):H surfaces after (i) a "HF-last" wet cleaning and (ii) a 1100 °C, 2-min hydrogen bake.

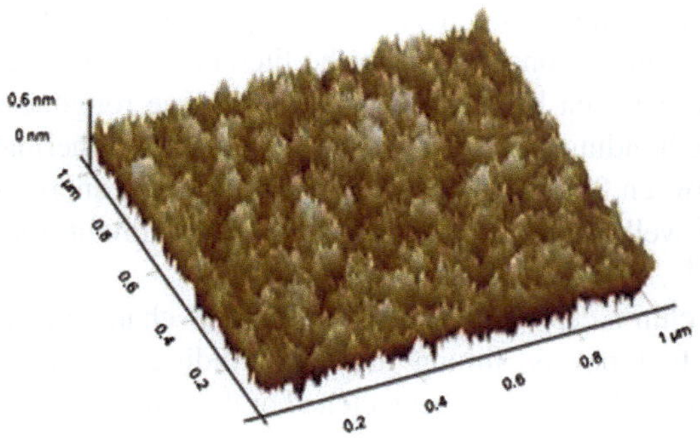

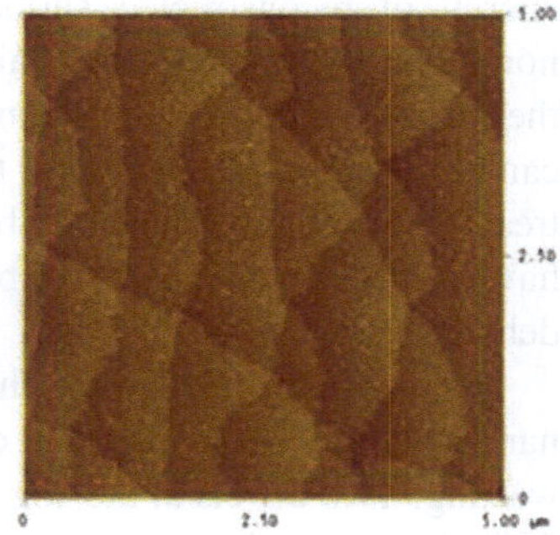

Fig. 4 AFM pictures of Si(001)H surfaces: **a** after a "HF-last" wet cleaning (1 × 1 μm^2 scan size). **b** after a 1100 °C, 2 min. H_2 bake (scan size: 5 × 5 μm^2); numerous wide terraces bordered by atomic step edges can be seen

This analysis has been made in the purpose of checking the freshly prepared surface. Images are provided in Fig. 3. The "HF-last" surface Z range (e.g., the height difference between the highest and the lowest point of the image) was 0.6 nm (scan 1 × 1 μm^2). No terrace was observed on that surface (Fig. 4a). Because of smoothing by the high-temperature H_2 bake, the Z range associated with the reconstructed surface is equal to 0.14 nm only, e.g., the height of a single atomic step. Large terraces are otherwise obvious in Fig. 4b. The terrace mean width is around 100 nm. The widest ones can reach 700 nm.

A well-ordered and smooth surface has thus been achieved; the next steps consist in a direct bonding and dicing steps in order to obtain the surface protection.

2.2.2 Bonding and Dicing

A direct bonding at room temperature has been applied in the clean room in order to reduce the contamination of the encapsulation. This step consists in putting into close contact a 200-mm silicon wafer which received the same reconstruction process described in the previous paragraph. The delay between reconstruction treatment and bonding must be the shortest possible in order to reduce the particles deposition. This time has been fixed at less than one hour for this study. The bonding energy keeping attached the two previous silicon wafers has been measured and compared with the energy of the non-reconstructed silicon wafers as reference. The bonding step has been optimized in order to obtain a low bonding energy providing an easy way to remove the silicon cap. However, at the opposite, a high bonding energy enabling a mechanic dicing without dies debonding is mandatory. A hydrophobic direct bonding step has been developed and the bonding energy has been measured to be between 110 and 140 mJ/m^2 while the reference bonding energy is around 20 mJ/m^2 at room temperature. Hence, the bonding energy of reconstructed surfaces is more than five times higher than that of non-reconstructed surfaces. This result is consistent with the literature and verifies the fact that a strong dependence between bonding energy and surface roughness can be observed [22]. Besides, the bonding energy can be increased when a thermal treatment is applied, typically between 200 and 300 °C [8]. Such bonding energies have been demonstrated to be well suited for dicing step, i.e., without dies debonding.

Moreover, to make easier the chip's cap removal in a UHV-STM chamber, the nanopackaged chips include etched bevels aiming also at avoiding substrates wasting. Two bevels at the edge of each chip have been etched at the dimensions of 500 µm wide and 50 µm deep. In order to define the dies dimensions for the dicing step, 1.3-mm-wide and 50-µm-deep etched lines have been also added. The nanopackaged chip has been designed in order to receive an organic or inorganic object too; consequently, cavities have been etched in the silicon cap with different sizes ranging between 200 and 2000 µm. Figure 5 represents infrared images of the bonded wafer including a half part of the wafer containing the cavities while the right part does not have it. Figure 5a represents the X-axis etched lines providing the right dicing blade positioning and 1-cm^2 die can be observed in Fig. 5b. The cavities have been well observed in the center of each etched lines in both figures.

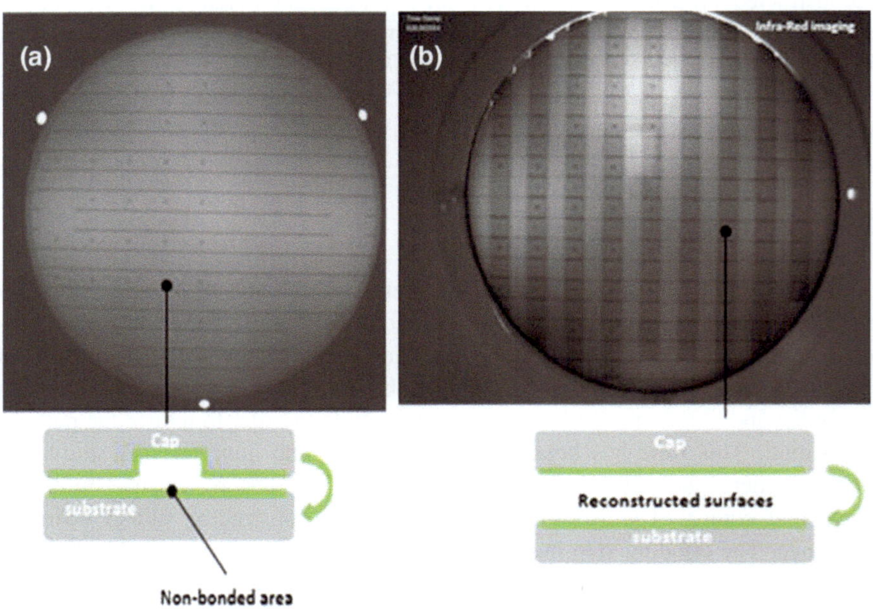

Fig. 5 Infrared images of a bonded SOI substrate with a silicon cap containing cavities in a half part of the 200-mm wafer. **a** X-axis guide etched lines for dicing step can be observed and **b** the 1-cm^2 chips after dicing step. The *left part* includes cavities in the center of the dies, the *right part* have been fabricated without cavities

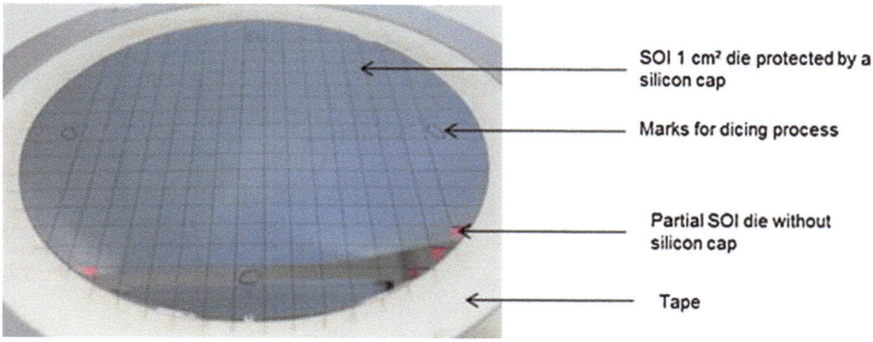

Fig. 6 Photograph of 200-mm SOI substrate bonded with a silicon cap. The wafer has been diced in 10×10 mm^2 dies. Regarding the number of 1-cm^2 die with a temporary silicon cap, a yield superior at 95% has been obtained

Following the guide etched lines, the 200-mm bonded wafers have been diced using a Disco dicing tool. Two steps dicing processes have been used: a S5550 dicing blade (150 μm wide) has been used for a cut depth of 615 μm. The dicing blade has been stopped before the bonding interface. A second run has been performed with a S4060 blade (110 μm wide) through the bonding interface up to the tape on which has been put the bonded wafers. The saw rotation speed has been fixed at 30 K revolutions/min (rpm). After an optimization process, no dies breakdown has been occurred and only several incomplete dies at the edge of the wafers have been detached as depicted in Fig. 6. The process repeatability has been established superior at 95% of yield considering the entire 1-cm^2 dies. The alignment of the bonded wafers during this step has been ensured by dicing marks previously etched during the process flow.

As represented in Fig. 7, nanopackaged dies have been obtained composed of a silicon or SOI substrate with a Si(001):H(2×1) surface, protected by a silicon temporary cap. The bevels can be observed at the edge of the chip, facilitating the opening of the temporary cap thanks to a blade insertion. The hermiticity is ensured

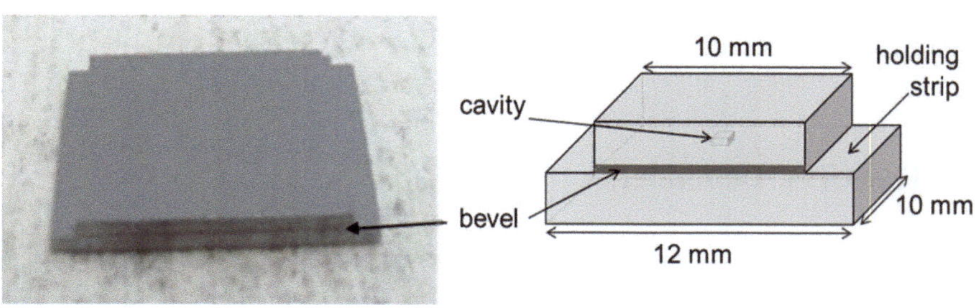

Fig. 7 Picture and scheme of a 12 mm × 10 mm nanopackaged chip composed of a Si(001):H (2×1) surface protected by a silicon temporary cap and including bevels in order to facilitate the opening of the die

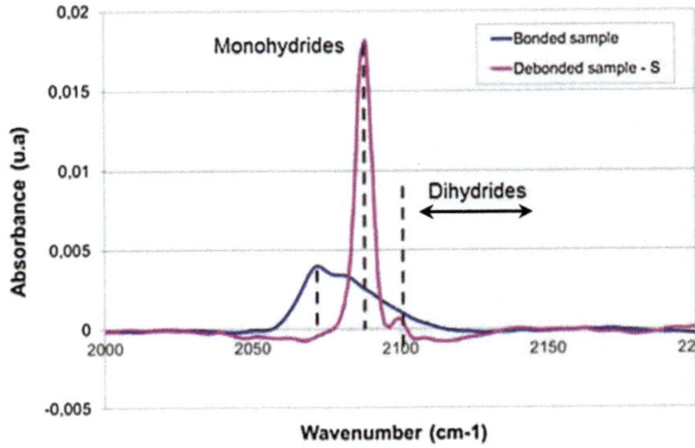

Fig. 8 Typical FTIR-MIR measurements of SiH peaks for "non-bonded" (2080 and 2095 cm^{-1}) and bonded (2065 and 2075 cm^{-1}) reconstructed surfaces

by the direct bonding of the cap and the surface is totally protected against the atmosphere contaminants. The sample can be stored during several months outside a clean room.

At the end of the process, a final control of the surface has been achieved on a die after silicon cap removal in order to verify the SiH surface saturation. A FTIR-MIR spectroscopy has been conducted in order to study the difference between a "bonded surface" and a free surface, i.e., none protected by a temporary cap. As depicted in Fig. 8, SiH peaks for "non-bonded" have been detected at 2080 and 2095 cm^{-1} and have been attributed to monohydrides [23, 24]. For bonded reconstructed surfaces, peaks (2065 and 2075 cm^{-1}) are shifted and are assumed to be due to monohydrides. Wave numbers of peaks attributable to dihydrides (2106 and 2116 cm^{-1}) are drawn for a better understanding.

FTIR-MIR spectrum shows significantly altered spectral shape: The peaks are shifted to lower frequency. This phenomenon has already been observed by Weldon et al. [25] for Si(111) reconstructed surface bonding. To explain this frequency shift, one should consider the modification of the monohydride environment due to the neighboring second wafer surface bonded.

To conclude this part, the 1-cm^2 chips have been successfully characterized and the results have demonstrated the presence of SiH saturation at the surface and large terraces after the reconstruction step achievement. These samples have thus been shipped to research laboratories in which atomic-scale surface studies have been performed using LT-UHV-STM systems.

3 The UHV Debonders

As presented in the previous paragraph, the Si(001):H(2×1) surface is protected by a temporary cap. A hydrophobic bonding [8] has been conducted, and the dicing step has been successfully achieved without cap debonding. A crucial result for the nanopackaging concept consists in controlling the opening of the temporary cap in the UHV chamber in order to have access to the chip surface for atomic-scale

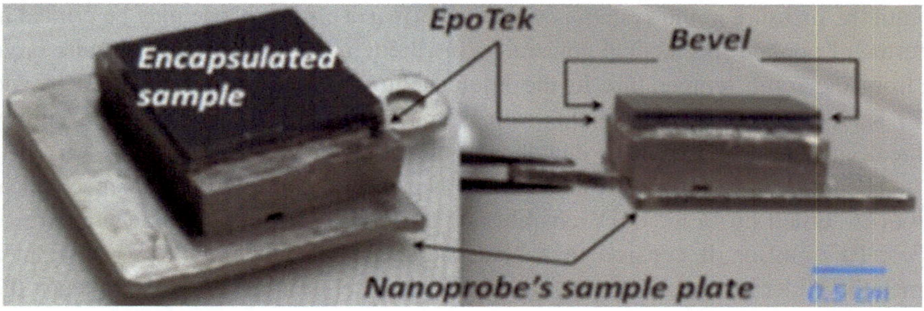

Fig. 9 A capped passivated silicon chip positioned on a Scienta Omicron sample plate using an EPO-TEK H21D glue. The blade of the UHV debonder will be introduced right inside one of the two bevels depending of the way the sample is pushed on the slider of the debonder

surface characterizations. The top-encapsulated passivated-silicon chips have to be opened under UHV conditions to avoid as much as possible any trace of contaminants before UHV-STM characterizations. As recalled in the above sections, Si(100)-H:(2×1) is a very reactive surface for the chemical contaminants in ambient conditions. Thus, we have chosen to minimize the number of native surface defects by opening the chip under UHV conditions. On a dice, the location of the bonding interface is indicated by a bevel (cf. Sect. 2.2.2) cut on the two opposite sides of the chip as illustrated in Fig. 9. To remove the cap, a robust cleaving blade is positioned flat as regards and toward the bevel. Then, the capped chip is pushed hard 500 μm deep inside the bevel between the cap and the sample. This toggles the cap out of the chip with a force applied corresponding to the force required at least to break the interfacial bonds between the two wafer-bonded substrates as detailed in the previous paragraph.

The selection of the glue to fix up robustly the backside of the silicon chip on a Scienta Omicron sample plate is an important step of the full process to be UHV and low-temperature compatible. Since in the following, the presented characterizations are performed on LT-UHV-STM systems, the selected glue has to stay conductive at 4 K, to be UHV compatible and hard enough to bear all the transferring steps through the successive chambers of our systems and first to resist the debonding step. Two kinds of glues have been tested: indium and a conductive epoxy polymer (EPO-TEK H21D). In our case, with our UHV "debonder" system EPO-TEK leads to the most reproducible results in term of stability. After preparing the EPO-TEK with a ratio 0.1/0.01 g and having left it at rest overnight, this preparation was applied as homogenous as possible on the sample transfer plate. Before gluing, the chip must be clean and non-oxidized at its backside to ensure a back ohmic contact with the metallic transfer plate. We have also taken care of the fact that the chip is perfectly free of any trace of constraints as cracks or pieces crumbling apart particularly at the edges of the dice. After positioning the chip on the glue at the surface of the transfer plate, the glue was polymerized by heating the sample up to 80 °C for 90 min in a standard laboratory oven. Finally, the encapsulated chip was inserted in the load-lock chamber of the LT-UHV-4 STM system until the pressure in the system is stabilized down. This last phase may be critical in terms of applied mechanical constraints, as the temperature gradient is quite large [from the room temperature to the liquid helium or nitrogen temperature (4 and 77 K, respectively)]. During this step, it may happen that some samples simply break into the UHV analysis chamber of our nanoprobe.

The first generation of UHV debonders is presented in Fig. 10a. This mechanical tool is settled up in an UHV chamber of our UHV system at a base pressure of 3×10^{-11} mBar. It consists in a blade adjustable in height with a screw, in a metallic slider where the encapsulated chip is inserted until it gently touches the blade. To push the cap away, a wobble-stick was used to abruptly push again the sample plate toward the blade but without touching the chip. It generally results in a fast ejection of the cap away of the debonder, within some cases a collision happens between the end of the debonder and the chip itself since the alignment of the wobble-stick end with the sample stage cannot be perfect in this case.

After several trials, this technique was proved to be too brutal for the chip as for the wobble-stick itself. Furthermore and already noticed above in our UHV system configuration, it is almost impossible for the wobble-stick end to hit always the

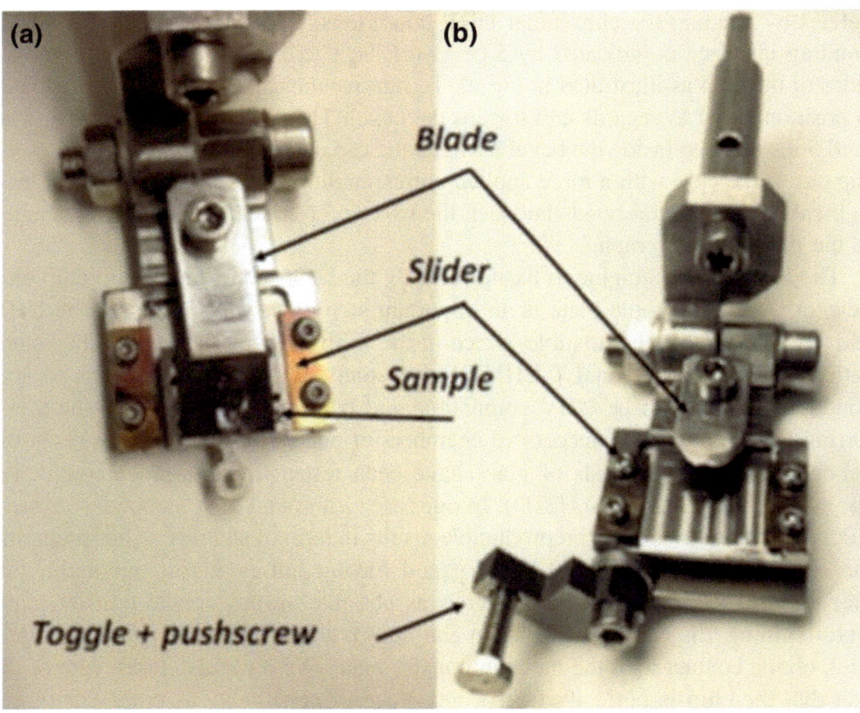

Fig. 10 The first (**a**) and second (**b**) generations of debonders. Different blade shapes have also been tried in order to minimize the constraints on the sample itself during the opening. The «push screw» on (**b**) is grabbed directly with the wobble-stick and rotated to push the sample toward the blade

same point on the trench of the sample plate when several trials are necessary to debond the cap. A second generation of UHV "debonder" was designed and fabricated as presented in Fig. 10b. A simple opening torque was added at the outer edge of the slider to hold the sample itself during the pushing toward the blade. This toggle owns an extra screw facing the sample which plays the role of a "push screw." After this toggle is latched down on the debonder, this push screw is directly grabbed by the end of a rotating wobble-stick. Then, the push screw is rotated, pushing the sample toward the blade until the ejection of the cap occurs out of the sample. This is the debonding system used for the results presented hereafter in this chapter. However, the rotating wobble-stick was shown to be too weak for some samples. A new generation of debonder is under study with a fixed sample and a slow step-by-step approach of the cleaving blade toward the bevel. These further modifications were thought to keep minimizing the mechanical constraints applied on the nanopackaged chip.

4 UHV Surface Characterizations After the Opening

After its opening, the sample is transferred into the next UHV chamber where the LT-UHV-4 STM is seating without any further preparation. Then, a first approach of the tungsten tips is performed using an optical camera to locate them through the window on the surface sample as presented in Fig. 11.

This step permits to take a large-scale glance at the surface and to visualize macroscopic defects, if any. The macroscopic defects appearing on the surface in Fig. 11 may come from the cap removal. They may also have been trapped at the bevel space, coming from the slicing of the 200 mm wafer in chips.

Thanks to the on-top UHV scanning electron microscope (SEM) standing just above the four STM tips of the LT-UHV-4 STM head and also above the sample, a careful look on the sample surface cleanness was most of the time performed first at room temperature with a resolution better than the above Fig. 11 optical inspection. Figure 12 SEM image is giving a general picture of a sample edge after its opening in the UHV. Very few damages seem to have been created with our debonding technique.

For the low-temperature STM characterization, the debonded Si(100)H chip was slowly transferred on the LT-UHV 4 STM stage maintained at 4 K to avoid a too fast change in temperature of the sample starting from room temperature. Therefore, the sample was left for at least 15 min on the shielding door of the LT-UHV 4 STM inner cryostat before its direct insertion on the 4 K LT-UHV 4 STM head. Finally, we have statistically STM-scanned large-size areas by positioning the tips of our LT-UHV-4 STM system on the Si(100)H surface using the viewport optical camera as presented in Fig. 12.

A first set of STM images was obtained at liquid nitrogen temperature (77 K). As a first example, Fig. 11 is indicating where Fig. 13 STM image presented below was recorded. Each STM scanner of our LT-UHV 4 STM is able to scan over a

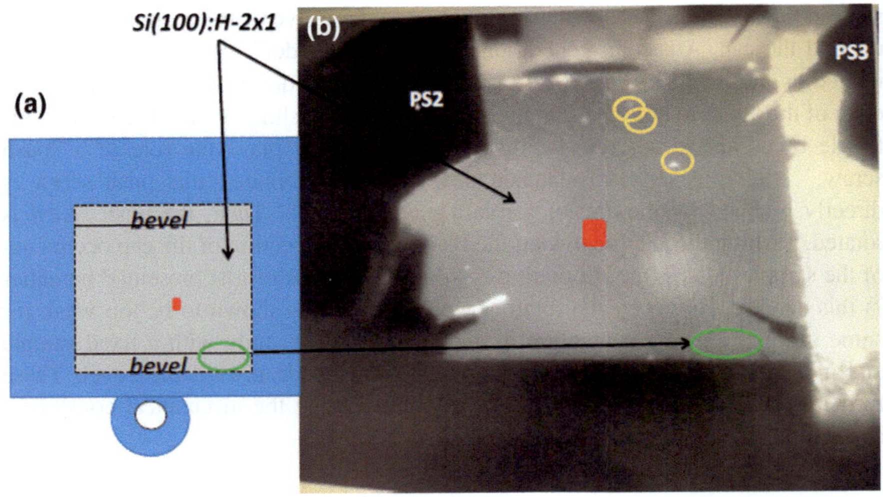

Fig. 11 **a** A schematic of a debonded sample (in *gray*) sitting on its sample plate (in *blue*). The *red* area is indicating where the tip loaded on the piezo scanner 3 (PS3-*right side* on **b**) is going to land on the surface. The location of the bevels where the blade was inserted to remove the sample cap is also indicated in A. The optical photograph **b** was recorded through a viewport of the LT-UHV-4 STM during the manual approach of the tips toward the surface of the chip. The 45° oriented tips loaded on PS3 and PS2 are reflecting on the Si(100)H surface. The area in *red* (**a**) was reported in (**b**) to persuade the reader that an STM tip apex can be very precisely located on the sample surface. The remaining trace of a bevel is indicated by the *green circle* on (**a**) and (**b**). The *yellow circles* are indicating the position of some macroscopic dusts on the surface

maximum window of about 2 μm × 2 μm. Notice that the complete chip surface is reachable by each scanner. But we have specifically imaged the Si(100)H surface area found under the tip apex of a given scanner. After a manual approach of the tip toward the surface, the Matrix software was left approaching gently and automatically a given tip by setting up the tunneling current intensity and the corresponding bias voltage like in a standard LT-UHV-STM. Here, we have used typically 2 V, 5 pA for the first approach in the tunneling regime. When the STM tip apex is stable enough, a large area can be scanned, typically in a scanning range between 50 and 500 nm (Fig. 13).

In average, we have found that the mono-atomic terraces are about 200–400 nm in lateral extension for the chips we have characterized and some mono-atomic stepped depletion areas are always observed distributed randomly all over a given terrace. Furthermore, the areas close to the edges of the Si(100)H chip are more defects covered than the central part of the chip as presented in Fig. 14.

Close to the edge of a given chip (red rectangle on the gray area Fig. 14b), it was possible to scan over some short mono-atomic terraces and their edges. Figure 14a empty-state STM image is characteristics of the relative disorder of this peculiar chip edge. Some SiH dimer rows can still be imaged running perpendicular from one terrace to another. Some dimers are decorated by some atomic-scale defects with defect density about 0.5 per nm^2 at the edge of this chip.

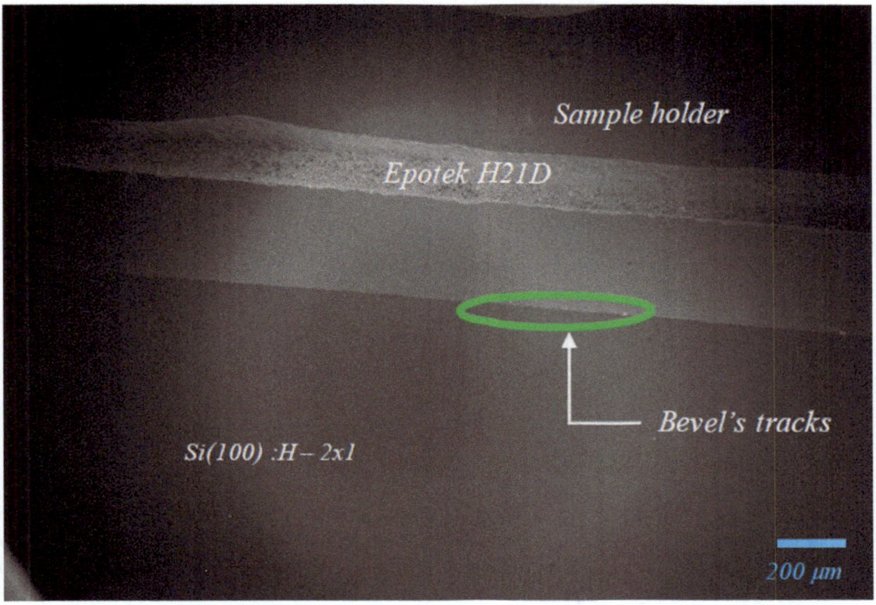

Fig. 12 An SEM image obtained at room temperature at the bevel-side edge of a debonded Si(100)H passivated surface. This is a quite uniform area without any kind of large damage even in the area where the blade hit the bevel. Similar images can be recorded at each edge and at all the surface indicating that apart from the large dusts observed in Fig. 11, this surface was not damaged after the debonding step

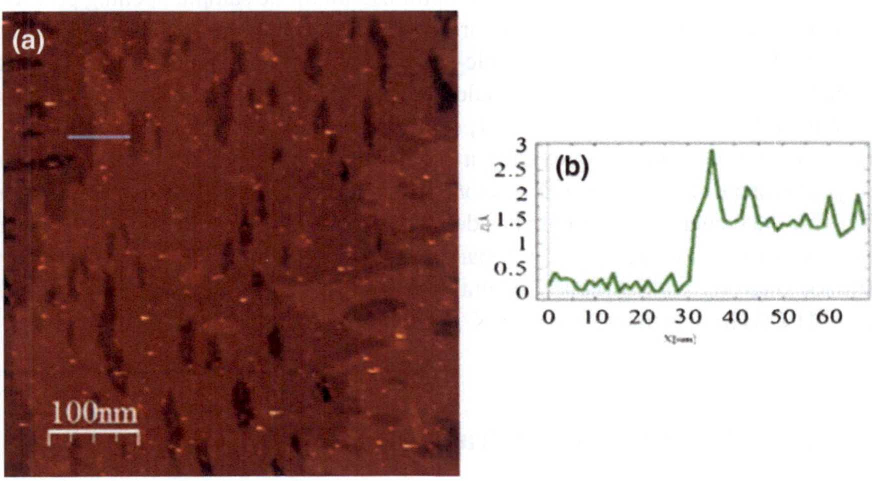

Fig. 13 **a** A large-scale constant current STM image (500 nm × 500 nm) of a UHV debonded Si(100)H surface recorded at 77 K using a vertically positioned tungsten tip with $V = 1.9$ V and $I = 5$ pA. The scanned area is relatively cleaned. There are large parallel more than 300 nm in width mono-atomic terraces running along this area of the surface. As presented in (**b**), the terrace STM image corrugation is below 50 pm with a characteristic mono-atomic step height for Si(100)H of about 150 pm. The large bumps on the top terrace as observed on (**b**) are due to very local atomic-scale impurities which remain to be identified

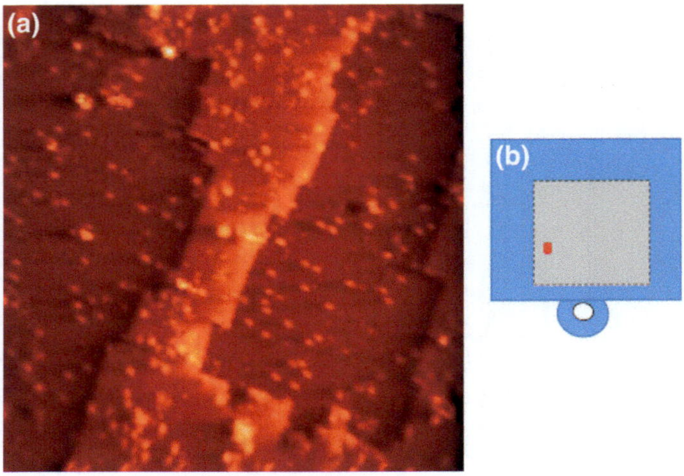

Fig. 14 **a** 50 nm × 50 nm constant current STM image recorded at 77 K on an area close to the edge of the Si(100)H debonded chip (scanning location indicated by the *red rectangle* on **b**) for 2.0 V and $I = 20$ pA. This image is indicating how the surface of this particular chip was not uniformly clean at its edge. In particular, it was difficult to scan over the multiple steps while maintaining a stable tip apex

Further investigations aim at understanding the origin of the defects and reducing the defect density. In order to reduce the particle level due to the dicing step of the chips, a chemical cleaning has been added in order to reduce the contamination during the opening. The optimization of the sample opening protocol is a key parameter and several tests are conducting which will enhance the defect density. Concerning the local atomic-scale impurities and mono-atomic stepped depletion areas, several studies could be started on the reconstruction surface treatment described in the Sect. 2.2.1, especially using a percent of Hcl gas during the reconstruction process providing a better smoothing and a lower incorporation of contaminant in the RP-CVD reactor. Nevertheless, the STM characterization at 77 K in the center of the sample has demonstrated a very high surface quality with flat and large Si(001):H terraces separated by single atomic steps. Consequently, surface analysis with a higher resolution has been successfully carried out with LT-UHV-STM at temperature of 4 K.

5 LT-UHV-STM Atomic-Scale Images

Atomic-scale images of the hydrogenated Si(001):H surface have been performed first with a single tip LT-UHV-STM after on the LT-UHV-4-STM whose head is equipped with 4 STM tip able to scan in parallel on the same surface.

5.1 LT-UHV-STM-1P Probe Surface Studies

The characterizations were carried out in a Krakow UHV system from Omicron NanoTechnology GmbH with a base pressure of 5×10^{-11} mbar. The UHV system consists of three interconnected chambers incorporating high-resolution SEM, LEED optics and low-temperature STM (LT-STM). All STM measurements were performed thanks to an Omicron LT-STM operating at the cryogenic temperature of 4 K with electrochemically etched polycrystalline tungsten tips used as probes. All values of bias voltage in STM experiments presented in the report are given with respect to the sample. SPIP and WSxM [26] software were used for image processing and STM data analysis.

For LT-UHV-STM characterization samples are transferred immediately after the debonding process to the LT-STM cryostat, where they are cooled down to 4 K. Because the samples prior to opening under UHV are protected by a bonded cap, we do not apply any additional preparation process. It is worth noting that this is a quite new approach in cooperation between industrial type of material processing site and the research laboratory. We are able to reach down lower temperature with the system such as liquid helium temperature (4.2 K). At a large scale, we are able to switch the tip bias voltage from positive (filled-state images) to negative one (empty-state images), chasing the differences between the charged defects, as dangling bonds, an atom or cluster laying on the surface as illustrated in Fig. 15.

Results are shown on the representative 100 nm × 100 nm filled-state images presented in Fig. 16. It has been confirmed that the surface crystallographic structure of the hydrogenated Si(001):H is relatively well preserved for large, flat, and single-terrace areas. The high-resolution STM nm-scale scans of the debonded

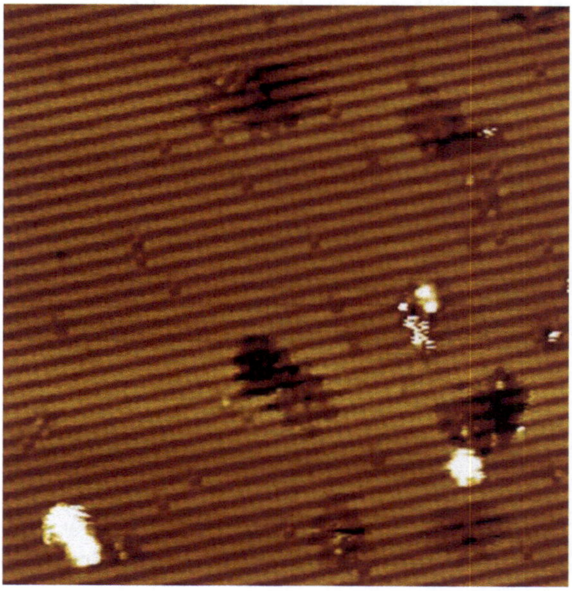

Fig. 15 High-resolution LT-STM image (STM scanning parameters: $V_{bias} = -2$ V, and $I = 10$ pA; 100 nm × 100 nm) of the Si(001):H surface after debonding of the sample no. H-dimer rows are clearly visible

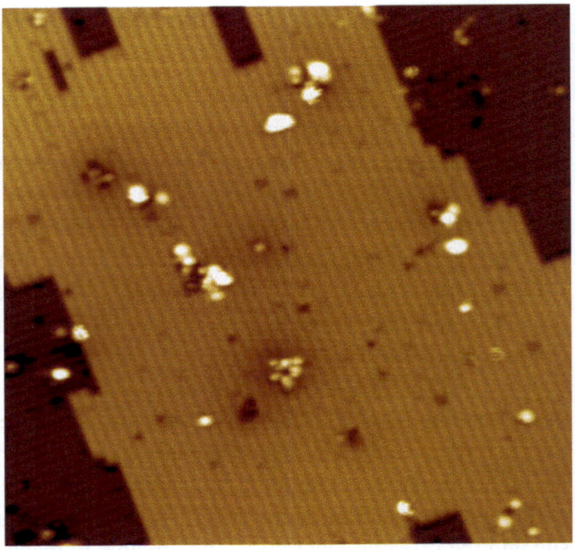

Fig. 16 High-resolution LT-STM image (STM scanning parameters: $V_{bias} = -2.5$ V, and I = 10 pA; 100 nm × 100 nm) of the Si(001):H surface after debonding of sample. H-dimer rows and atomic step edges are clearly visible

sample clearly show that the surface is passivated by a monolayer of hydrogen atoms, which form a stable monohydride (2 × 1) phase in which each Si surface atom is bound to one H atom. The agglomeration of defects (white corrugations) observed on the atomically clean terraces in both figures could be due to the chemical processing used during the surface preparation or the wafer debonding procedure itself. In fact, Si vacancy agglomerations with the similar structures and density (black contrast) indicate that the defects could be formed during the debonding process and white corrugations consist of the Si adatom agglomerates. The average defect densities estimated from several STM scans on the debonded surfaces are about 1 defect per 40–50 nm^2. The average defect densities estimated from several STM scans on the debonded surfaces are about 1 defect per 50 nm^2. Even if these defects have been reported and the density has been consistent with the literature [27], further investigations are running to discriminate the surface reconstruction impact and the debonding effect. To conclude, the reconstructed Si(001)-(2×1):H surfaces have been successfully observed at the atomic scale and can be used for nanodevice fabrication.

Another characteristic feature of the analyzed sample structure is lack of any dangling bond defects. The overall high quality of the surface allows for surface DB structures construction by hydrogen atoms desorption with the use of a STM tip.

5.2 *LT-UHV-4 STM Surface Studies*

Another Si(100)-H:(2×1) surface study at the atomic resolution has been conducted using a more recent STM tool built by Scienta Omicron GmbH and containing 4

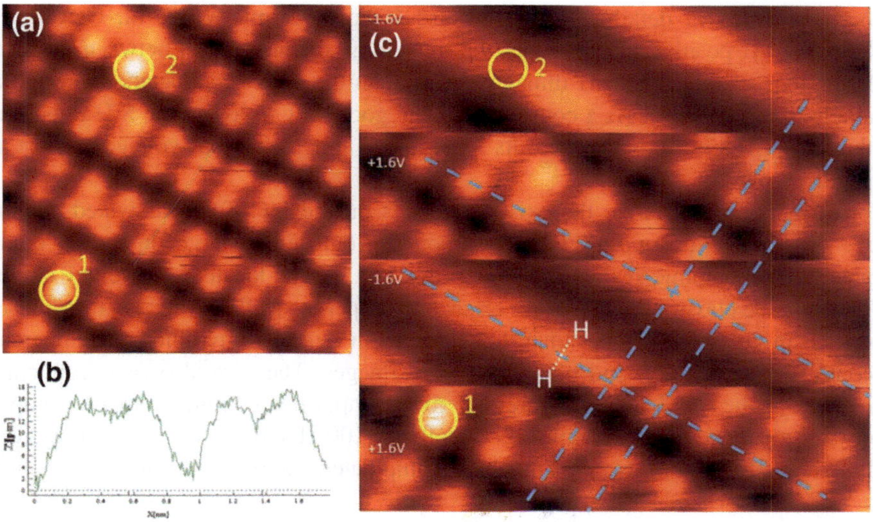

Fig. 17 Atomically resolved constant LT-UHV-STM images (**a**, **b**) recorded with scanner 4 of the LT-UHV-4STM head at small scanning range (4 nm × 4 nm) using a vertical tip configuration. The *left image* in (**a**) was recorded at + 1.6 V for $I = 35$ pA. The (**b**) image was recorded alternatively shifting the bias voltage between +1.6 V and −1.6 V during one image recording with the sequence indicated on (**b**). The *blue dotted line* is indicating the positions of the SiH dimers parallel and perpendicular to the 2 × 1 reconstruction of the Si(100)H surface. The *yellow-circled* defects are underlined to prove that (**a**) and (**b**) are exactly the same scanned areas. **c** is giving a scan line extracted from (**a**) +1.6 V image with a 15-pm corrugation. On **b**, the position of the two hydrogen atoms of the same SiH dimer is represented

piezoelectric scanners. The whole UHV-4 STM chamber was maintained at a base pressure of 3 × 10^{-11} mBar during this study. After having located rather free of defects large-terrace areas on the chip, the LT-UHV-4-STM was cooled down to 4.2 K keeping the Si(100)H debonded sample on its sample stage. At 4.2 K, the Toulouse PicoLab (CEMES) LT-UHV-4 STM had already demonstrated a Δz noise level better than 2 pm on an Au(111) surface [28]. Figure 17a is presenting a small 4 nm × 4 nm area constant current image of the small portion of Si(100)H surface selected in the center of the chip. This is a typical positive bias voltage image of the 2 × 1 monohydride rows reconstruction. A scan line was extracted from this image perpendicular to the raw (Fig. 17c) confirming the nice 2 pm noise level of our LT-UHV-4-STM.

Care must be taken to interpret Fig. 17a STM image in the prospect of determining the exact atomic-scale structure of this imaged portion of the Si(100)H surface. For this purpose and to explore further the surface contrast variation as a function of the bias voltage, a systematic change of the bias voltage was performed switching alternatively the bias voltage between +1.6 and −1.6 V during the recording of a constant current STM image. As presented in Fig. 17b and at −1.6 V, the maximum STM contrast perpendicular to the rows is exactly located in between the 2 small maxima obtained at +1.6 V. This half-row apparent shift has been analyzed in detail in [29] (see also Chapter Atomic Wires on Ge(001):H Surface). At positive bias voltage, the tunneling conductance in between the two hydrogens of the same SiH dimer is larger than the direct conductance through a single SiH bond [29]. As a consequence, the good way to locate the SiH dimer on the Si(100)H surface is to record its STM image at negative bias voltage and not at positive.

To better characterize the existing defects on the Si(100)H passivated silicon surface after its debonding, Fig. 18 is pointing out defects larger in size as compared to the 2 atomic-scale defects imaged in Fig. 17 and located by a yellow circle. They can be classified by size and electronic properties [30]. In Fig. 18, the defects circled in green are smaller than those circled in blue and give the same apparent contrast in the empty- and filled-state STM images. They are also easy to manipulate with the STM tip. They are superficial impurities coming from the sample environment and are not chemisorbed on the Si(100)H surface. On the contrary, the blue-circled impurities cannot be manipulated during scanning but are

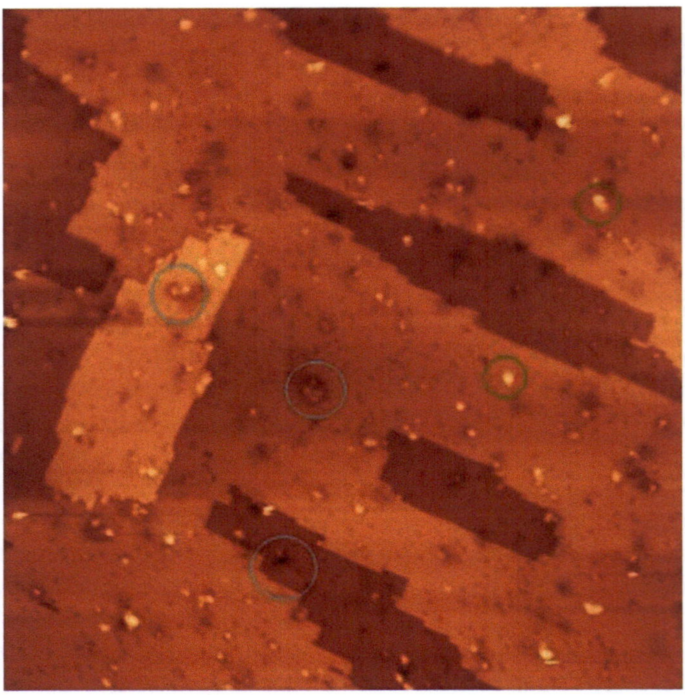

Fig. 18 A large 100 nm × 100 nm constant current STM images recorded in the center of the debonded Si(100)H chip at 4.2 K for −1.67 V and I = 15 pA. Several nanoscale surface defects have been *circle* in *blue* and *green*. The *blue-circled* defects are surrounded by a native dark hallow indicative of a LDOS-depleted areas induced by the impurity itself. The green-circled defects can normally be manipulated by the STM tip during a scan. Scanned areas as presented in this figure are clean enough not to contaminate the tip. On this debonded chip, those are the best places to start constructing atomic-scale circuits

electronically unstable by changing the bias voltage. In the filled-state images, they are generally surrounded by a dark shadow (Fig. 18) indicating a LDOS depletion around such a defect, a signature of an acceptor-like type of defect. It can be residual native doping species surfacing from the bulk during the thermal treatment of the fabrication process or simply some residual oxygen remaining in the chamber. Large-size defects with an apparent diameter of 10 nm or more may be coming from surrounding dusts deposited on the surface during the debonding. They are quite rare over the surface of the imaged chips.

Coming back to atomic-scale defects, we have observed that the chip surface is covered by a few native Si single dangling bonds (DBs) and also by dihydride SiH2 defects that can be observed both in filled- and empty-state images as presented in Fig. 19. For filled-state images, each DB is appearing as a bright protrusion along a dimer row with generally the DB protrusion being shifted off the SiH dimer center as presented in Fig. 19. In empty-state images, a single DB is spotted as a dark depletion and can be imaged surrounded by a dark halo [31] depending on the tip, the doping level of the silicon substrate and on the bias voltage. Bare silicon dimers and dihydride dimers can also be easily imaged on the Si(100)H surface as presented in Fig. 19a, b. Those are STM imaged as bright asymmetric protrusions

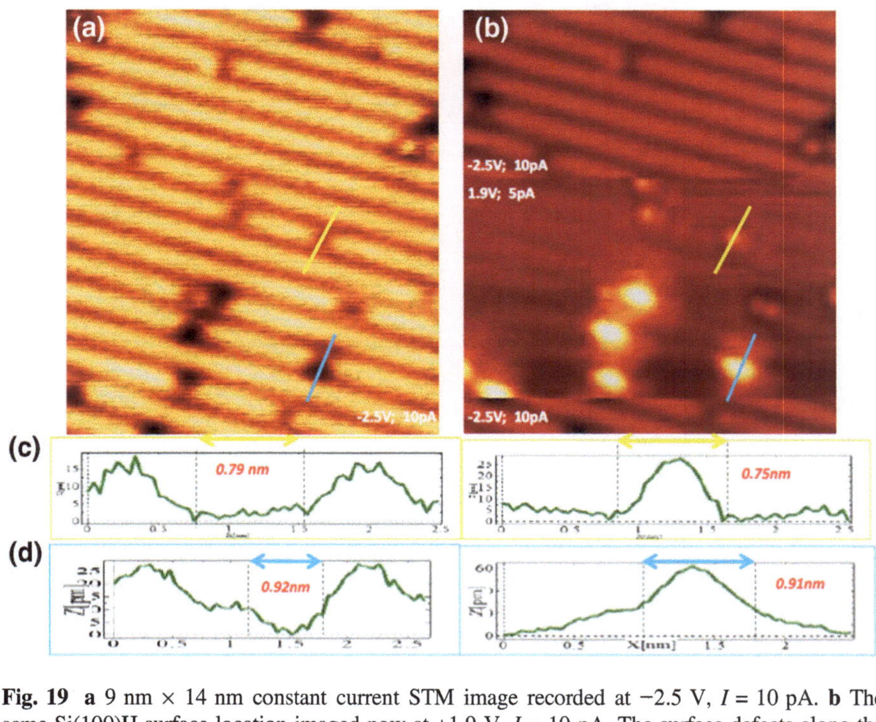

Fig. 19 **a** 9 nm × 14 nm constant current STM image recorded at −2.5 V, I = 10 pA. **b** The same Si(100)H surface location imaged now at +1.9 V, I = 10 pA. The surface defects along the dimers rows appearing as a depletion for negative bias voltage and as large protrusion for positive one. Scan lines were extracted in both cases as presented in (**c**) and (**d**) (*yellow* and *blue lines* in **a** and **b**, respectively). Corrugation of these images is about 60 pm maximum

along the dimer row for empty-state images and as a dark depletion in the filled-state images. The asymmetry in the STM images is due to the buckling of the silicon dimers.

Finally, returning on a clean Si(100)H surface area, single *I–V* curves were recorded to measure the surface apparent electronic gap relative to the bulk Fermi level position which must be located in the middle of the valence band–conduction band gap since the used silicon wafer is intrinsic. *I–V* curves can be recorded on many location on this Si(100)H surface showing different tunneling conductance for the same feedback current. In the case presented in Fig. 20, the STM tip apex was positioned above a hydrogen atom of a single DB dimer marked in Fig. 20A by a white star. Our LT-UHV-4 STM head being stable enough for long time single shot *I–V* recording, the feedback loop was turned off during more than a tenth of seconds to ramp the bias voltage starting from −1.6 V up to +2.0 V and back to record the consequent tunneling current intensity. After the recording of the final *I–V* point, the feedback loop is turned on again to recontrol the tip apex height relative to the surface. Following this process, the apparent Si(100)H surface energy

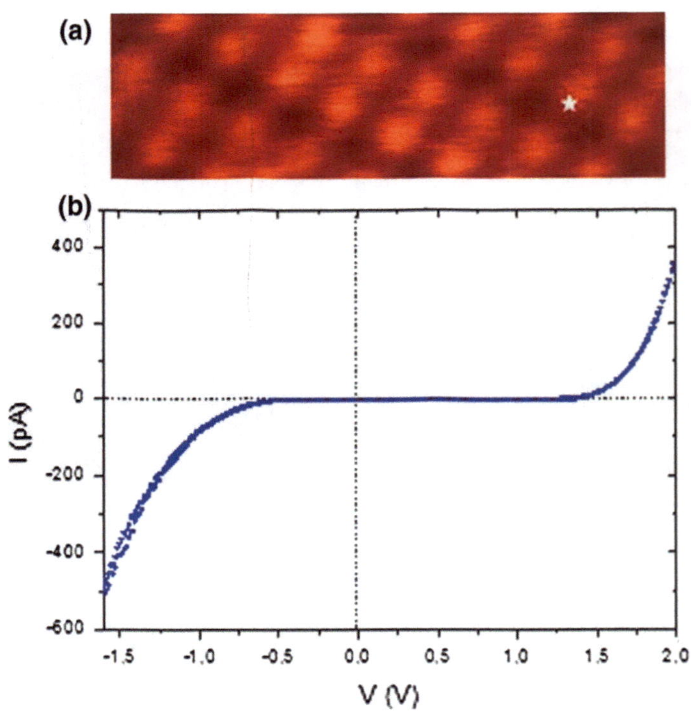

Fig. 20 **a** 4 nm × 1.3 nm constant current STM image recorded at 4.2 K for +1.6 V and I = 35 pA. **b** A one shot I–V curve recorded on a single hydrogen atom location for a setup current of 35 pA with a bias voltage scanned from −1.6 V to +2.0 V both with the feedback loop off during this recording. The apparent gap is about 1.7 eV for this initial +1.6 V and I = 35 pA tip apex to surface relative distance

band gap was found to be about 1.7 eV. This value is smaller than the known 2.1 eV surface gap for the Si(100)H surface because Fig. 20 *I–V* characteristics was recorded in the z direction (sample grounded) meaning with a true k_z direction 1.1 eV electronic bulk silicon gap. The surface 2.1 eV gap in the k_x (k_y) gap direction of the surface band structure can only be measured using a floating 2 STM tip surface measurement, i.e., with the back sample ground disconnected. Depending on the setup feedback current, this apparent 1.7 eV gap can vary from larger than 2.1 eV to its minimum of 1.1 eV.

6 Summary and Conclusion

In conclusion, this chapter clearly demonstrated the interest of a nanopackaging approach for the UHV environment applications. A technology has been developed on 200-mm Si or SOI wafer in a standard industrial-like clean-room way enabling to obtain a hydrogenated Si(001):H-(2×1) reconstructed surface, protected by a temporary silicon cap. After opening, the surface can be used without applying any additional preparation process usually applied in a UHV chamber requiring a complex protocol. This surface preparation is composed of a chemical cleaning and a hydrogen silicon surface passivation which provides a low particle density and a well-ordered reconstructed surface. FTIR characterization has demonstrated a full SiH monohydrides surface passivation of the silicon surface. A hydrophobic direct bonding between SOI substrate and a silicon wafer has been conducted and which protects the surface against contamination due to the atmospheric air exchange such as hydrocarbons or particles. The bonded wafers are diced into 1-cm^2 samples in order to be loaded in a STM holder. The nanopackaged samples can thus be stored during several months before its opening and use by the research laboratory. To remove the cap, a robust cleaving blade must be positioned at the bonding interface in a UHV chamber. The samples are transferred immediately after the debonding process to the UHV- LT-STM chamber and then on its sample stage, where they are cooled down to 4 K.

An atomic resolution has been obtained and demonstrated that the surface crystallographic structure of the hydrogenated Si(001):H is relatively well preserved for large-, flat-, and single-terrace areas. The high-resolution STM scans at the atomic scale have clearly shown that the surface is passivated by a monolayer of hydrogen atoms, which forms a stable monohydride (2 × 1) phase and in which each Si surface atom is bound to one hydrogen atom. Agglomeration of defects has been observed on the atomically clean terraces, with a higher density in the edge of the samples. However, the quality of the surface is convenient in the major part of the surface for molecule grafting or surface nanoscale device fabrication.

Further investigations are under way in order to improve the surface, focusing on reconstruction step or chemical cleaning step on the die. These process changes should positively improve the Si(001):H surface quality in the foreseeable future. Others works concern the surface studies after a dopant implantation. In order to connect the nano-object on an implanted area, the doped surface damages must be removed by the reconstruction treatment while carefully controlling the dopant diffusion effects. These product developments have demonstrated the fabrication of such structures on an industrial scale and directly usable for atomic electronic devices or technologies based on molecular logic circuits thanks to the collaboration between clean room microelectronics players and research laboratories.

Acknowledgements This research has been supported by the 7th Framework Program of the European Union Collaborative Project ICT (Information and Communication Technologies) "Atomic Scale and Single Molecule Logic Gate Technologies" (ATMOL), contract number: FP7-270028. The CEA Institute, the CEMES, and the Polish National Science Center are acknowledged as well for financial support.

References

1. Morris, J.E.: Nanopackaging: Nanotechnologies and Electronics Packaging, p. 1. Springer (2008)
2. Joachim, C., Gimzewski, J.K., Schlittler, R.R., Chavy, C.: Electronic transparence of a single C60 molecule. Phys. Rev. Lett. **74**(11), 2102–2105 (1995)
3. Lafferentz, L., Ample, F., Yu, H., Hecht, S., Joachim, C., Grill, L.: Conductance of a single conjugated polymer as a continuous function of its length. Science **323**(5918), 1193-1197 (2009)
4. Ohnishi, H., Kondo Y., Takayanagi, K.: Quantized conductance through individual rows of suspended gold atoms. Nature **395**(6704), 780–783 (1998)
5. Thuaire, A., et al.: Nanopackaging: from Nanomaterials to the Atomic Scale, Advances in Atom and Single Molecule Machines vol. 7, p. 113. Springer Series (2015)
6. Lwin, M.H.T., Tun, T.N., Kim, H.H., Kajen, R.S., Chandrasekhar, N., Joachim, C.: Silicon on insulator nanoscale backside interconnects for atomic and molecular scale circuits. J. Vac. Sci. Technol. B **28**(5), 978-984 (2010)
7. Reynaud, P., et al.: Nanopackaging solution from clean room to UHV Environment: Hydrogen Passivated Si (100) Substrate fabrication and use for atomic scale investigations and self-assembled monolayer grafting. Procedia Eng. **141**, 121–129 (2016)
8. Rauer, C., Moriceau, H., Rieutord, F., Hartmann, J.M., Fournel, F. Charvet, A.M., Bernier, N., Rochat, N., Dansas, H., Mariolle, D., Morales, C.: Mechanism involved in direct hydrophobic Si(100)-2$\times$1:H bonding. Microsyst. Technol. **21**(5), 961-968 (2015)
9. Kolmer, M., Godlewski, S., Zuzak, R., Wojtaszek, M., Rauer, C., Thuaire, A., Hartmann, J.M., Moriceau, H., Joachim, C., Szymonski, M.: Atomic scale fabrication of dangling bond structures on hydrogen passivated Si(100) wafers processed and nanopackaged in a clean room environement. Appl. Surf. Sci. **288**, 83–89 (2014)
10. Weber, B., Mahapatra, S., Ryu, H., Lee, S., Fuhrer, A., Reusch, T.C.G., Thompson, D.L., Lee, W.C.T., Klimeck, G., Hollenberg, L.C.L., Simmons, M.Y.: From Ohm's law survives to the atomic scale. Science **335**(6064), 64–67 (2012)
11. Fournel, F., et al: Accurate control of the misorientation angles in direct wafer bonding. Appl. Phys. Lett. **80**(5) (2002)
12. Le Gac, G., Audoit, G., Thuaire, A., Moriceau, H., Baillin, X.: Nanovias FIB-etching and filling in a micro-nano interposer for molecular electronics. In: Technical Proceedings of the 2012 NSTI Nanotechnology Conference and Expo, NSTI-Nanotech **2**, 539 (2012)
13. Jerier, P., Dutartre, D.: Boron autodoping in single-wafer epitaxy of silicon at reduced pressure. J. Electrochem. Soc. **146**(1), 331–335 (1999)

14. Ishikawa, Y., Kumezawa, M., Nuryadi, R., Tabe, M.: Effect of patterning on thermal agglomeration of ultrathin silicon-on-insulator layer. Appl. Surf. Sci. **90**(1–4), 11–15 (2002)
15. Jahan, C., Faynot, O., Tosti, L., Hartmann, J.-M.: Agglomeration control during the selective epitaxial growth of Si raised sources and drains on ultra-thin silicon-on-insulator substrates. J. Cryst. Growth **280**(3–4), 530–538 (2005)
16. Abbadie, A., Hartmann, J.-M., Holliger, P., Semeria, M.N., Besson, P., Gentile, P.: Low thermal budget surface preparation of Si and SiGe. Appl. Surf. Sci. **225**(1–4), 256–266 (2004)
17. Bühler, J., Steiner, F.-P., Baltes, H.: Silicon dioxide sacrificial layer etching in surface micromachining. J. Micromech. Microeng. **7**(1), R1–R13 (1997)
18. Meyerson, B.S., Himpsel, F.J., Uram, K.J.: Bistable conditions for low-temperature silicon epitaxy. Appl. Phys. Lett. **57**(10), 1034–1036 (1990)
19. Trucks, G.W., Raghavachari, K., Higaschi, G.S., Chabal, Y.J.: Mechanism of HF etching of silicon surfaces: a theoretical understanding of hydrogen passivation. Phys. Rev. Lett. **65**(4), 504–507 (1990)
20. Wostyn, K., Rondas, D., Kenis, K., Loo, R., Hikavyy, A., Douhard, B., Mertens, P.W., Holsteyns, F., De Gendt, S., D'Urzo L., Van Autryve, L.: Use of a purged FOUP to improve H-terminated silicon surface stability prior to epitaxial growth. ECS Trans. **64**(6), 669–673 (2014)
21. Hersam, M.C., Guisinger, N.P., Lyding, J.W., Thomson, D.S., Moore, J.S.: Atomic-level study of the robustness of the Si (100)-2×1: H surface following exposure to ambient conditions. Appl. Phys. Lett. **78**(7), 886–888 (2001)
22. Rieutord, F., Capello, L., Beneyton, R., Morales, C., Charvet, A.M., Moriceau, H.: Rough surface adhesion mechanisms for wafer bonding. Electrochem. Soc. Trans. **3**, 205–215 (2006)
23. Chabal, Y.J., Patel, C.K.N.: Solid hydrogen in amorphous silicon: phase transition. Phys. Rev. Lett. **53**(18), 1771–1774 (1984)
24. Chabal, Y.J.: High-resolution infrared spectroscopy of adsorbates on semiconductor surfaces: Hydrogen on Si(100) and Ge(100). Surf. Sci. **168**, 594–608 (1986)
25. Weldon, M., et al.: Surface Science **368**(1–3), 163–178 (1996)
26. Horcas, I., Fernández, R., Gómez-Rodríguez, J.M., Colchero, J., Gómez-Herrero, J., Baro, A. M.: WSXM: a software for scanning probe microscopy and a tool for nanotechnology. Rev. Sci. Instrum. **78**, 013705 (2007)
27. Huang, D.H., Yamamoto, Y.: Atomic manipulation of hydrogen on hydrogen-terminated silicon surfaces with scanning tunnelling microscope. Force Microsc./Scan. Tunneling Microsc. **3**(4), 50–62 (2002)
28. Yang, J., Sordes, D., Kolmer, M., Martrou, D., Joachim, C.: Imaging, single atom contact and single atom manipulations at low temperature using the new ScientaOmicron LT-UHV-4 STM. Eur. Phys. J. Appl. Phys. **73**(1), 10702 (2016)
29. Yap, T.L., Kawai, H., Neucheva, O.A., Wee, A.T.S., Troadec, C., Saeys, M., Joachim, C.: Si (100)-2×1-H dimer rows contrast inversion in low-temperature scanning tunneling microscope images. Surf. Sci. **632**, L13-L17 (2015)
30. Buehler, E.J., Boland, J.J.: Identification and characterization of a novel silicon hydride species on the Si(100) surface. Surf. Sci. **425**(1), L363-L368 (1999)
31. Kawai, H., Neucheva, O., Yap, T.L., Joachim, C., Saeys, M.: Electronic characterization of a single dangling bond on n- and p-type Si(001)-(2×1):H. Surf. Sci. **645**, 88–92 (2016)

Atomic Wires on Ge(001):H Surface

Marek Kolmer, Jakub Lis and Marek Szymoński

Abstract The drive toward miniaturization of electronic devices motivates investigations of atomic structures at semiconductor surfaces. In this chapter, we describe a full protocol of formation of atomic wires on Ge(001):H-(2×1) surface. The wires are composed of bare germanium dimers possessing dangling bonds, which introduce electronic states within the Ge(001):H surface band gap. With a view to the possible applications, we present detailed analysis of the electronic properties of short DB dimer lines and discuss strong electron–phonon coupling observed in STM experiments on single DB dimers. For longer DB dimer wires, this coupling is attenuated making their usage in future nanoelectronic devices feasible.

1 Introduction

In recent years, numerous nanometer-scale systems created with atomic precision have been subjected to detailed experiments and simulations (see [1] and references therein) for the purpose of implementation of novel electronic devices. Particularly, the hydrogen-passivated (001) surfaces of germanium and silicon are promising candidates for future nanoelectronic applications. The main reason behind such a large interest in the hydrogenated Ge(001):H and Si(001):H surfaces is the unique possibility of creation of predesigned surface dangling bond (DB) structures by the scanning tunneling microscopy (STM) tip-induced desorption of hydrogen atoms [2, 3]. This STM-based lithography can be atomically precise. It makes it possible to tune the position of electronic states by suitable DB (defect) arrangement which is a cornerstone of the successful development of novel atomic-scale logic circuits [4] and atomic interconnects [5].

M. Kolmer (✉) · J. Lis · M. Szymoński
Faculty of Physics Astronomy and Applied Computer Science, Centre for Nanometer-Scale Science and Advanced Materials, NANOSAM, Jagiellonian University, Lojasiewicza 11, 30-348 Krakow, Poland
e-mail: marek.kolmer@uj.edu.pl

M. Kolmer and C. Joachim (eds.), *On-Surface Atomic Wires and Logic Gates*, Advances in Atom and Single Molecule Machines,
DOI 10.1007/978-3-319-51847-3_3

In the electronic applications, the transport properties of defined DB structures play a key role. A lot of effort was put to gain the experience needed to design the device architecture for a given functionality. In these theoretical works, the vibrational excitations have been disregarded, though the experimental data demonstrate their relevance for systems consisting of DBs [6, 7]. Although the phenomena can be investigated with the means of the STM, the problem has been up to now rarely addressed.

In this chapter, we present our recent efforts in understanding of properties of atomic-scale DB structures on the Ge(001):H surface by means of STM-based experimental methods. We start from a protocol for ultra-high vacuum (UHV) preparation of the hydrogenated Ge(001):H surface and a short description of theoretical methods useful in modeling of STM experiments on that surface (Sects. 2 and 3). In Sect. 4, we describe basic electronic properties of the Ge(001):H surface related particularly to the observed Fermi level pinning. In the next section, we present two efficient protocols for STM tip-induced hydrogen desorption from the Ge(001):H surface which allow the formation of predesigned and complex DB structures. Due to much lower bang gap of Ge(001):H surface, we demonstrate that, unlike at its silicon counterpart, the electronic properties of STM-fabricated DB structures can be directly explored by STS d*I*/d*V* technique. In Sect. 6, we present detailed experimental STS d*I*/d*V* analysis of short DB dimer wires, which are also supported by transmission coefficient *T*(*E*) calculations [2]. Finally in Sect. 7, we discuss STM current-induced dynamical behavior of DB structures comprised of DB dimers. Particularly, flipping events of a DB dimer between two energetically equivalent geometries allow us to explain the symmetry of the empty-state STM image of the isolated DB dimer [6, 7]. We summarize the chapter with perspectives related to direct planar electron transport experiments performed on long, atomically defined DB wires.

2 Surface Preparation and Experimental Methods

Our experiments are carried out in an ultra-high vacuum. The STM measurements are performed with the ScientaOmicron low-temperature scanning probe microscope (LT-STM/AFM) and low-temperature 4-probe STM system [8]. The base pressure of both systems is below 1×10^{-10} mbar. The preparation chambers are equipped with a noble gas ion gun, homebuilt hydrogen cracker and an infrared pyrometer. The LT-STM imaging is carried out in both systems at reduced temperature of around 4 K (liquid helium) with etched tungsten tips used as probes.

The preparation of the Ge(001) surface starts with a wafer (TBL Kelpin Crystals, 0.5 mm thickness, undoped) dicing into $2 \times 10\ \text{mm}^2$ samples. Next, the samples are put inside an ultrasonic washer and cleaned in acetone (10 min, 310 K) and ethanol (10 min, 300 K). Finally, the samples are mounted on sample holders and introduced to UHV system. The samples are first outgassed for 8 h at 500 K. Then we anneal the samples by direct current heating at 800 K until the pressure in the preparation chamber does not exceeds low 10^{-10} mbar level. The main part of sample preparation protocol consists of cycles of 1 keV Ar^+ sputtering for 15 min with the sample kept at 1040 K. Typically after 3–5 such cycles, an atomically clean and reconstructed Ge(001) surface is obtained as presented in Fig. 1a. The microscale quality of surface reconstruction is also monitored by a low-energy electron diffraction (LEED) technique. The main difference between above UHV preparation protocol and the one used for Si(001) surface lies in the Ar^+ sputtering process, which is not needed for silicon preparation. This difference comes mainly from a lower melting point of germanium ($\sim$1211 K) as compared to silicon ($\sim$1687 K), which prevents Ge annealing at high temperatures.

Fig. 1 Bare Ge(001) surface after the UHV preparation procedure. **a** Large 100×90 nm^2 STM image obtained at LHe temperature (4.5 K, −2 V, 100 pA). **b** Different reconstructions of a Ge (001) surface observed on STM images at LHe temperature (*left* and *central*) and at room temperature (*right*)

The preparation of hydrogenated Ge(001):H surfaces is performed by the use of homebuilt hydrogen cracker composed of an UHV leak valve attached to a molecular hydrogen line and a tungsten filament cooled by a water shroud. The bare Ge(001) surface is exposed to atomic hydrogen formed by a precisely dosed H_2 flux coming through the hot tungsten filament. During hydrogen passivation procedure, the sample is kept at 485 K and the hydrogen partial pressure in the UHV chamber is maintained at 1×10^{-7} mbar. The above routine results in hydrogenated Ge(001):H surface with monohydride (2 × 1) reconstruction consisting of germanium dimer rows with each surface Ge atom possessing one H atom attached (Fig. 2). There are about 2–4% defective surface unit cells which translates to the defect density of the order of 4×10^{12} cm^{-2} mostly in the form of native bare Ge atoms (single atoms or dimers, Fig. 2). The STM constant height topography images strongly depend on the STM bias, and particularly, these two types of native

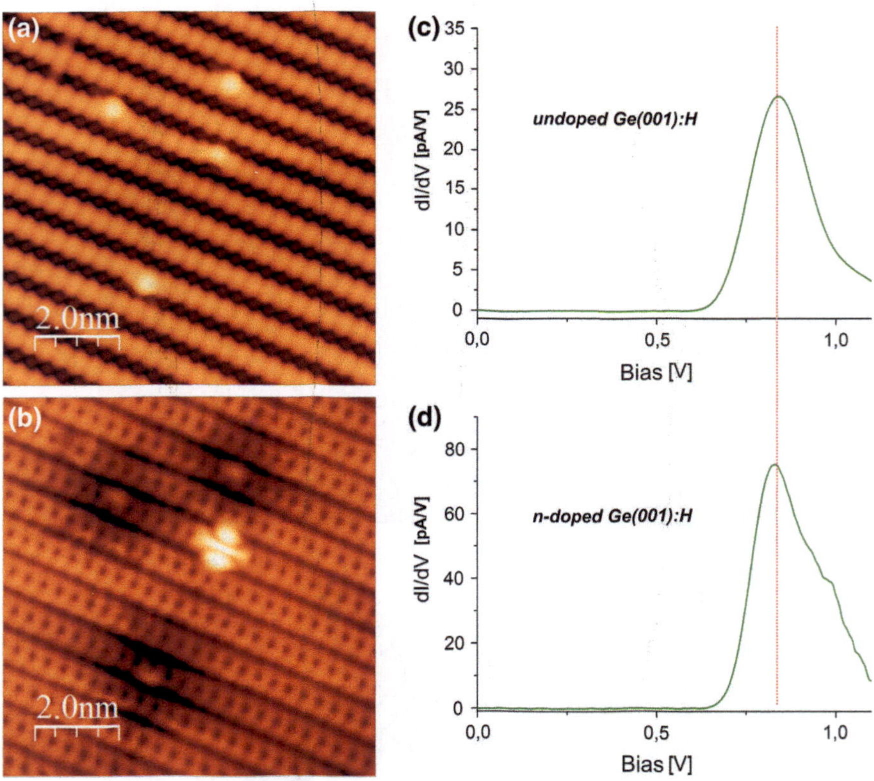

Fig. 2 Filled **a**- and empty **b**-state STM images of the Ge:H surface with three single DBs and one DB dimer. The images are acquired with 20 pA tunneling current (reprinted with permission from [6]). **c**, **d** STS d*I*/d*V* data acquired over the DB dimer on **c** undoped (n-type, ~45 Ωcm) and **d** highly n-doped (n-type, Sb-doped, 0.008–0.009 Ωcm) samples

defects can be easily recognized on empty-state STM images (Fig. 2b). A single unsaturated Ge atom (single DB) is surrounded by dark halo resulting from its negative charge state, whereas the bare Ge dimer (DB dimer) is observed as a characteristic "butterfly" pattern [6, 7]. In this chapter, we focus only on DB structures consisting of DB dimers and particularly we discuss the mechanism behind the formation of the "butterfly."

3 Calculation Methods

At present, the methods based on Density Functional Theory (DFT) are the first choice computational tool in condensed matter and surface physics and chemistry. They allow calculations of the ground state properties like the lattice constant, the geometric configuration and phonon frequencies as well as the electronic density. It is also a common and successful practice to consider the self-consistent eigenfunctions calculated within the DFT schemes as the physical wave functions. The hybrid functional (HSE06) offers the description of the germanium that closely coincides with the experimental findings [9]. But, as shown in the latter paper, for many systems the generalized gradient approximation performs satisfactorily. The investigation of surface properties of the germanium is a tricky issue; the crystal perturbations appear to attenuate on longer distances than for silicon. A lesson to learn from the discussion about the character of the top of the valence band at the Ge(001) surface [10] is that the results need to be checked against the slab width. For slabs too thin (6 layers), the band structure is not credible. Also, the displacements of atoms from their bulk positions are observed as deep as in the seventh layer beneath the surface. A minimal number of layers needed are nine, but even 12 layers cannot be termed eccentric. In the paper [11], it is argued that the interaction between the modeled surface and the bottom of the slab may be minimized for nonstandard bottom preparation: Instead of hydrogen termination, the surface structure is modeled on both sides (plate model). Then, all atoms in the slab are allowed to be relaxed.

The simple Tersoff–Hamann approximation for STM images works well for the Ge(001) surfaces discussed here. This technique allows a qualitative comparison between experiment and theory. To model dI/dV spectra or STM images recovering the experimental surface corrugation, more sophisticated tools are needed, e.g., surface Green-function matching (SGFM) method (refer to [2] and see the references therein). These methods describing qualitatively a full STM junction are used in Sect. 5 to model STS dI/dV spectra obtained experimentally on short DB dimer wires.

4 General Properties of Ge(001) Surface: Fermi Level Pinning

There are two low-temperature reconstructions of the bare Ge(001) surface: p(2×2) and c(4×2) (Fig. 1b). At higher temperatures, (2 × 1) reconstruction is also observed (Fig. 1b), but it is an averaged image due to dynamical changes to the surface geometry, and we will not discuss this reconstruction below. Upon cleavage the (001) surface, two unsaturated bonds per surface germanium are created. Their number is halved by forming bonds between two adjacent surface atoms. This pairing results in the appearance of long wires comprised of the atomic pairs (dimers). Further lowering of the surface energy is achieved by alternative tilting of the adjacent dimers along the wire. The buckling changes the electronic structure. "Already in 1979 Chadi proposed that the surface dimers of Si and Ge(001) could buckle (...). The buckling opens a gap between occupied and unoccupied surface states lowering the energy of the dimer. Chadi used a tight-binding scheme to optimize the total energy of the dimerized surface without imposing any symmetry constraints. He found that the surface dimers spontaneously buckle, and that the symmetric dimer configuration is unstable. Buckling of the dimers lowers the energy of the occupied states and increases the energy of the unoccupied states." [12] As the atoms get nonequivalent, the electron density concentrates on the upper atom. The hybridization of the lower atom becomes closer to sp^2 than to sp^3 (which is the bulk hybridization) [13]. The pairs on neighboring reconstruction rows seem to be nearly noninteracting: the two reconstructions, i.e., c(4×2) and p(2×2), result from two possible arrangements of buckled dimers in the adjacent rows. Energetically, these two reconstructions are very similar, and it is possible to switch between them with STM applying the appropriate scanning parameters [14]. The surface electronic structure has been elucidated from scanning tunneling probe microscopy and angle-resolved photoelectron spectroscopy experiments; see, e.g., [15]. The lower surface band, called usually D_{up}, is located very close to the top of the bulk valence band. The electric neutrality of the surface requires that this band be fully occupied. There is an upper surface band about 0.3–0.45 eV above the lower band called D_{down}. It is located within the bulk band gap which for germanium reads 0.742 eV at 0 K and 0.661 eV at room temperature.

The unoccupied surface states form quasi one-dimensional bands as evidenced by experimental and numerical studies [16, 17] showing that the band is highly anisotropic with much larger dispersion along the rows (about 0.6 eV) than in the direction perpendicular to them. Such conclusions were drawn from the careful observations of the evanescent waves at surface steps. There was a heated debate about the occupied surface states D_{up} [10, 16]. It was concluded that these states with the electronic density concentrated at upper atom of the dimers spread along the direction parallel to the rows. In the perpendicular direction, there are actually no surface states but surface resonances.

The distinction between the surface and bulk character of the states can be made by comparison of spectra of clear and hydrogen-terminated spectra obtained in photoelectron spectroscopy experiments. In this way, the bulk character of what could be considered as the top of the surface valence band was demonstrated [16]. The hydrogen-terminated Ge(001):H surface is chemically inert, its highest occupied states have the energy deep below the valence band maximum (3 eV) [18, 19], and the energy of the lowest unoccupied state is above the bulk conduction band minimum. So, passivation washes away all surface states from the neighborhood of the band gap. The surface band gap measured with STM is broader than the bulk one and reads about 0.8 eV. The structure of the Ge(001):H surface closely resembles its silicon counterpart. The hydrogen termination consists in attaching one hydrogen atom to every surface germanium atom. The row structure stays untouched in the process. The chemical bonds between the atoms forming the dimers are not broken, but the symmetry between the dimer atoms is restored (Fig. 2). From our viewpoint, this surface is crucial. The single rows created upon selective hydrogen atom desorption geometrically and electronically recover characteristics of the rows at the bare surface.

The above facts embedded in the theory of the surface Fermi level pinning rationalize the following statements. For the bare Ge(001) surface, the surface Fermi level should be located below D_{down}. For the Ge(001):H surface, the surface Fermi level is supposed to be equal to the bulk one for the absence of the surface states to accumulate the electric charge. The experimental results are surprising, though. In both cases, the surface Fermi level is pinned to the top of the valence band. There is a difference: For the bare surface the Fermi level appears to be slightly above the maximum, while for the passivated one the SFL is below the maximum of the valence band. Consequently, for Ge(001):H it is possible to repetitively obtain STM images for all voltage biases, even within the band gap. This is impossible for the bare surface where at slight positive biases (within the surface band gap) such attempts end up with the tip crush. The difference may reflect the change to the actual surface Fermi level upon passivation, but it may follow from the chemical activity of the bare surface preventing close approach of the tip to the surface.

A metallic character of the bare Ge(001) surface has been known for a long time [20]. It hardly can be understood within the surface electronic structure outlined above. But, what cannot be reconciled with the conventional picture is the insensitivity of the surface Fermi level to the saturation of dangling bonds with hydrogen. There are several reports showing this phenomenon with different techniques [18, 19, 21]. One can imagine that some charge accumulates at the bare surface, but it hardly can be justified that similar surface charge is accumulated after passivation. In [21], it is reported that the surface Fermi level at Ge(001):H is slightly below the top of the valence band, no matter if n-type doped or p-type doped samples are investigated. The phenomenon is also illustrated in Fig. 2c, d. It shows STS spectra DB dimer defect in the passivation layer recorded for differently doped samples. The STS spectra are unaltered for these two samples. It is proposed that the dopant migration in a subsurface layer is at the origin of the phenomenon but its detailed mechanism is still not well understood. This phenomenon has mesoscopic/ microscopic manifestations. Note that for n-type doped samples, there is an inversion layer close to the surface; i.e., the holes are the majority carriers in a subsurface layer. As a result, the supplied current likely explores the thin subsurface layer only and two-dimensional current flows can be expected. Indeed, in the electric current transport experiments performed at room temperature in the four-point probe geometry, a qualitative difference between the n-type doped and the p-type doped samples was observed [22]. It turns out that for n-type doped samples the voltage at the surface is well described by function ~ln(r), where r is the distance from the source or drain. This voltage drop is characteristic to the two-dimensional systems. For the p-type doped samples, the voltage drop obeys 1/r low which points out to the three-dimensional character of the current flow [21].

5 Construction of Dangling Bond Wires on Ge(100):H Surface

In this section, we present two efficient protocols for atomically controlled STM tip-induced H extraction at liquid helium temperature conditions. In both cases, we start from a perfectly hydrogenated Ge(001):H surface area (Fig. 3a).

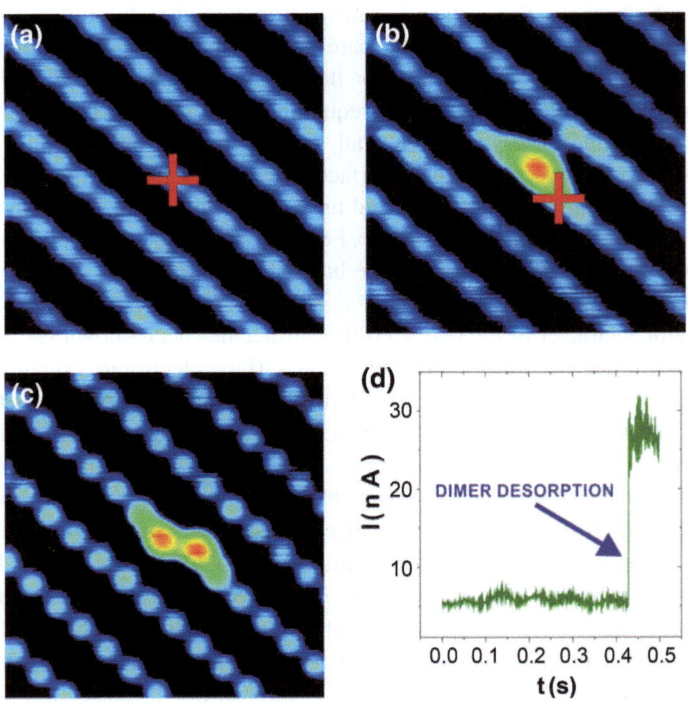

Fig. 3 STM tip-induced fabrication of a DB line running along the surface reconstruction rows. Red crosses indicate the positions of the tip during the dimer desorption processes, **a** filled-state STM image (−0.5 V, 1 nA, 4 nm × 4 nm) of the hydrogenated Ge(001) surface before desorption, **b** single DB dimer, **c** two neighboring DB dimers forming the DB line, **d** typical I (t) characteristic recorded during the desorption process (reprinted with permission from [2])

In the first protocol, the extraction is performed by pulsing the STM tip bias voltage with the STM feedback loop open. In this case, we approach the tip over the hydrogen dimer selected for extraction with the STM feedback loop parameters set on about I = 1 nA tunneling current intensity and a V = − 0.5 V bias voltage. The tip apex is positioned over the dimer according to the Ge(001):H filled-state STM image (see Fig. 3a). Subsequently, the feedback loop is turned off and the desorption process starts with a voltage pulse set up to positive biases, higher than +1.5 V (the exact value depends on the exact STM apex and typically is not larger than +2 V). The desorption of the hydrogen dimer is detected when a sudden rise of the tunneling current is observed in the $I(t)$ characteristic. The procedure can be repeated step-by-step until the targeted DB dimer pattern is constructed. The above protocol allows for the efficient construction of a predesigned DB nanostructure with atom-by-atom precision. Figure 3 illustrates an atomically controlled dimer-by-dimer desorption leading to the construction of a short 2 DB dimer line parallel to the Ge(001):H rows. The example of longer DB dimer wire oriented across surface reconstruction rows formed in the same way is presented in Fig. 4a. Note that here, unlike in the case of the Si(001):H surface, our STM tip V pulse protocol extracts a pair of H atoms per pulse instead of a single H. This H extraction method is, however, time-consuming, and so it is limited to DB structures containing up to tens of DB dimers.

To improve the efficiency of DB structure formation, we propose another method which uses the STM feedback loop to control the tunneling current constant during the desorption process and we move the STM tip along a defined path above the surface. Starting again with perfectly hydrogenated surface area, we place the

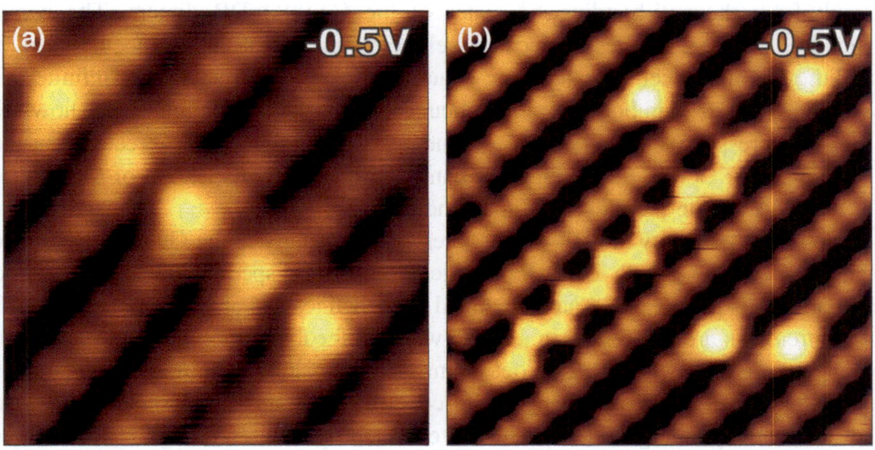

Fig. 4 Atomically perfect DB dimer wires obtained by two different STM-based hydrogen desorption protocols. **a** Five DB dimer wire oriented across Ge(001):H reconstruction rows ($3.5 \times 3.5\ nm^2$, 1 nA) formed by dimer-by-dimer hydrogen desorption. **b** Twelve DB dimer wire oriented along Ge(001):H reconstruction rows formed by closed feedback line desorption of hydrogen atoms (*courtesy* R. Zuzak)

STM tip above the center of hydrogenated germanium dimer. Then we change the STM feedback loop to conditions allowing very efficient H desorption. The exact STM parameters depend again on the particular STM tip apex, but tunneling current typically is in the range of 2–3 nA and bias is about +2.5 V. Then we move the apex along surface dimer rows at a constant speed ($\sim$5 nm/s). To reduce the possibility of drift-induced errors in the resulting DB line, we do not extend the pattern more than 10 nm. In most cases, the newly formed DB structure possesses still some single H atoms, which can be then removed by the previous protocol. The example of atomically perfect 12 DB dimer wire manufactured by the above method is presented in Fig. 4b. In this case, we use only one feedback closed loop line desorption procedure which is followed by two single H atoms desorption events induced by the feedback open method.

6 Electronic Properties of Short DB Wires[1]

Having already established the methods of wire formation, we can analyze their electronic properties by both dI/dV STS spectroscopy experiments and $T(E)$ calculations [2]. First to characterize the electronic properties of the Ge(001):H surface and of the bare Ge(001) surface, their electronic band structures are calculated (Fig. 5). The fully hydrogenated Ge(001):H surface is predicted to have a 1.1 eV surface electronic band gap, while the bare Ge(001) surface has a 0.6 eV surface band gap (Fig. 5a, d, respectively). The Ge(001):H surface gap decreases when DB dimer lines are created on the surface due to the electronic states introduced by the DB dimers near the bottom of the Ge(001):H conduction band edge. These bands result from the anti-bonding π^* states of the Ge(001) DB dimers. The corresponding bonding π states are located well below the top of the valence band edge, and they do not affect the size of the band gap. The band structures for infinite DB lines on Ge(001):H in both the perpendicular and the parallel direction are shown in Fig. 5b, c, respectively. In both cases, a new π^* conduction band is created near the bottom of the Ge(001):H conduction band. A significant dispersion of 0.67 eV is only found for the parallel DB dimer line (Fig. 5c).

To investigate how the DB dimer states shift gradually in the Ge(001):H surface band gap as the length of DB line increases, dI/dV spectra are measured and transmission spectra, T(E), are calculated for DB lines of various lengths. All STS measurements are performed in a mode with the feedback loop turned on between every two $I(V)$ characteristics to determine the tip position. The $I(V)$ characteristics are automatically collected using a grid covering a 2.5 nm $\times$ 2.5 nm surface area, and the corresponding dI/dV spectra are obtained by differentiating the $I(V)$ curves averaged previously over the area of the DBs only. The dI/dV spectra are simulated by calculating the electronic transmission spectra through the tunnel junction used

[1](Excerpt reprinted from [2] with minor editorial changes)

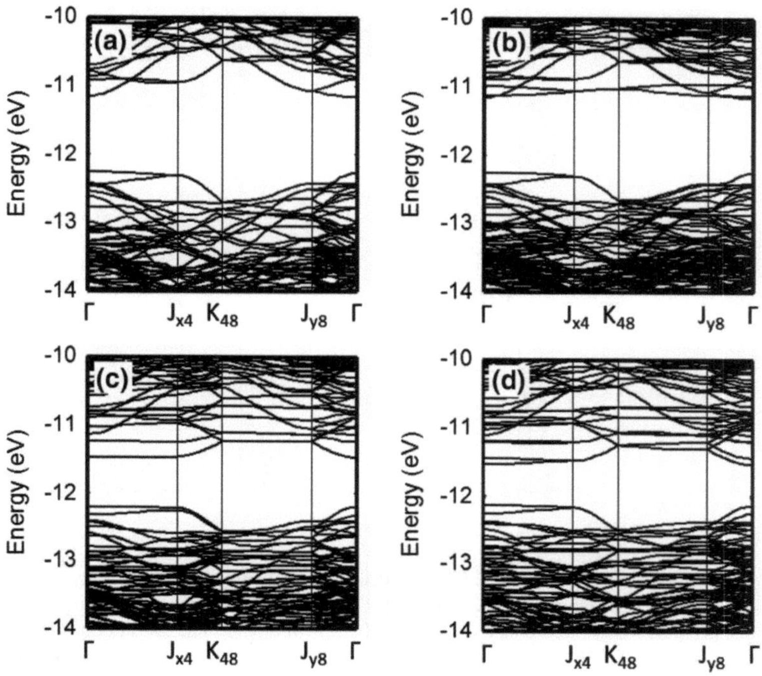

Fig. 5 DB structures of **a** a fully passivated Ge(001)-(2×1):H surface, **b** infinite DB line perpendicular to the dimer rows with 7 H-passivated dimers between two DB lines, **c** infinite DB line parallel to the dimer rows, where the DB dimers are buckled out-of-phase along the dimer row and **d** a clean Ge(001)-c(4×2) surface. A 9-layer slab is used to model the Ge surfaces. (Reprinted with permission from [2])

for the constant current image calculations, which consists of the tungsten tip, the Ge tip apex, the Ge(001):H surface and the Ge(001) bulk. The tip apex was placed 7 Å above the Ge(001):H surface.

First, the d*I*/d*V* spectra for the bare Ge(001) surface and for the fully hydrogenated Ge(001):H surface are measured and compared with calculated *T*(*E*) spectra. The experimental d*I*/d*V* spectra clearly show that the surface band gap increases upon surface hydrogenation from 0.25 eV for the bare Ge(001) surface to about 0.85 eV for Ge(001):H (light blue and green curves in Fig. 6a, respectively). The calculated *T*(*E*) spectra (Fig. 6b) follow the same trend, in agreement with the calculated bare Ge(001)-c(4×2) and Ge(001):H surface band structures (Fig. 5a, d, respectively). The band gap for d*I*/d*V* and *T*(*E*) spectra differs slightly from the calculated band structures, because a 9-layer slab is used for the band structure calculations while a semi-infinite structure is used in the transport calculations. Note that *T*(*E*) is calculated for a single point instead of averaging over the DB area. Therefore, the relative heights of the *T*(*E*) resonance peaks are different from the experimental d*I*/d*V* spectra.

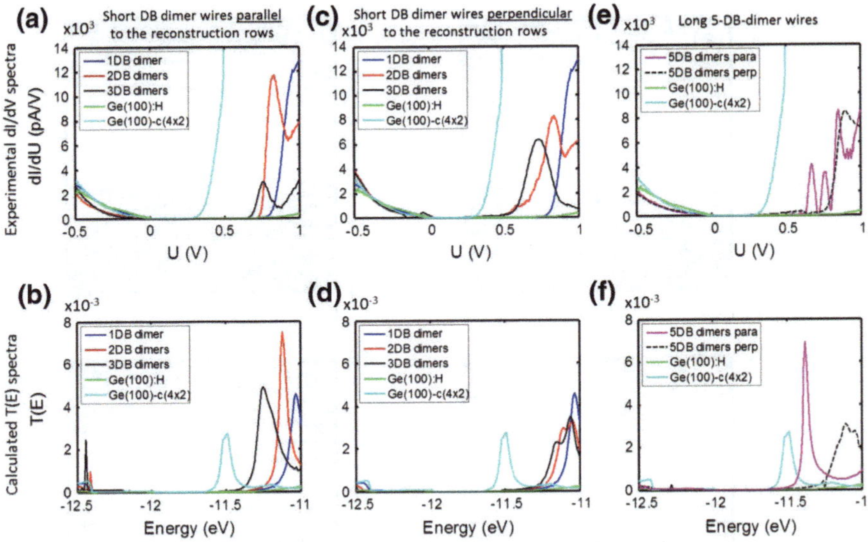

Fig. 6 **a**, **c**, **e** Experimental d*I*/d*V* and **b**, **d**, **f** calculated *T*(*E*) spectra for a fully hydrogenated Ge (001)-(2×1):H, DB lines containing 1, 2, 3 and 5 DB dimers aligned parallel and perpendicular to the Ge(001):H dimer rows and clean Ge(001) surface with c(4×2) reconstruction (STS feedback loop at 0.5 nA and −0.5 V for all cases except for 2 and 3 DB dimers in (**a**) and 5 DB dimers in (**e**), where 0.3 nA and −0.5 V set-point is used) (reprinted with permission from [2])

Next, experimental d*I*/d*V* spectra and calculated *T*(*E*) spectra are compared for DB lines with 1, 2, 3 and 5 DB dimers in both the perpendicular and the parallel direction. For DB lines in both directions, the experimental d*I*/d*V* spectra and the calculated *T*(*E*) spectra show large nonzero conductances at energies below the Ge (001):H surface conduction band edge (Fig. 6). Note that a large resonance peak appears 0.9 eV above the Fermi level even for a single DB dimer (blue resonance, Fig. 6a, c), clearly showing the DB dimer state below the conduction band edge. The position of the single DB dimer resonance peak is also described well by the calculations (blue resonance peak, Fig. 6b, d). No resonance peaks are observed near the valence band edge.

For DB lines parallel to the dimer rows, it is expected that the d*I*/d*V* resonances are observed below the conduction band edge and that the resonances will gradually span the 0.6 eV energy difference between the Ge(001):H and the bare Ge(001) conduction band edge as the length of the DB line increases. This is because each DB dimer introduces an additional π* state near the bottom edge of the Ge(001):H conduction band, and an infinitely long DB dimer line results in a dispersive π* band as shown in Fig. 5c. For short DB lines composed of 2 and 3 dimers, the d*I*/d*V* resonances indeed gradually shift toward lower energies compared to the resonance peak for a single DB dimer (Fig. 6a, red and black resonances), which is also observed in the calculated *T*(*E*) spectra (Fig. 6a, red and black resonances). When the number of DB dimers increases to 5, the d*I*/d*V* resonances shift further to lower energies (Fig. 6e, pink resonance), gradually approaching the conduction band edge of the bare Ge(001) surface (Fig. 6e, light blue resonance). The *T*(*E*) spectra show the same trend (Fig. 6d). For DB lines parallel to the dimer rows, the gradual shift of the resonance peaks results from the significant electronic couplings between the nearest-neighbor DB dimer states. Note that only one resonance is observed in the *T*(*E*) spectrum for the 5 DB dimer line, whereas a few peaks are observed in the d*I*/d*V* plot. This is because the tip position is fixed above the central DB dimer for the *T*(*E*) calculation whereas the d*I*/d*V* plot is averaged over the DB line. Since each *T*(*E*) peak results from a different DB dimer along the DB line, different peaks are enhanced in *T*(*E*) depending on the tip position. In all cases, no significant shift is observed in the valence band edge.

Similar to DB lines parallel to the dimer rows, the resonance peaks shift to lower energies for lines of 2 and 3 DB dimers perpendicular to the dimer rows. However, the shift in the measured d*I*/d*V* spectra is larger than the shift in the calculated *T*(*E*) spectra (Fig. 6c, d). For perpendicular DB lines, the shift in the *T*(*E*) resonances due to coupling between DB states is expected to be small because the band structure for an infinite perpendicular DB line shows only a nondispersive π^* band located at the edge of Ge(001):H conduction band (Fig. 5b). This difference between the measured and calculated shifts, however, results from a competition between surface atomic structure relaxation toward their infinite configurations and inter-dimer electronic interactions along those DB lines. Since the inter-dimer distance is larger for a perpendicular DB line, the buckling of the dimers for short 2 and 3 DB dimer lines is more flexible than for short parallel DB lines. For example, when the buckling of the perpendicular DB dimers is reduced by 25%, the DB-derived states shift down by almost 0.1 eV and show a resonance shift similar to the one observed in the d*I*/d*V* spectra (Fig. 6c). This flexibility causes the d*I*/d*V* shifts for short perpendicular DB lines to be similar to those observed for short parallel DB lines. However, when the length of the perpendicular DB line exceeds 3 DB dimers, the buckling becomes less flexible and approaches the buckling for the infinite DB line. Therefore, when the number of DB dimers is increased to 5, the resonance peak in both the d*I*/d*V* and the *T*(*E*) spectra shifts up in energy and becomes close to the position for a single DB dimer (dashed black resonance, Fig. 6e, f). This behavior is very different from the trend observed for the parallel DB line with 5 DB dimers.

7 Excitations of Short DB Wires: Electron–Phonon Interaction

As we will show, there is a strong coupling between the DB unoccupied state and phonons meaning that, upon charging, the geometry of the DB dimer defect substantially changes. Below, we describe some theoretical aspects of tunneling through the quantum dot strongly coupled to vibrations. We assume the Born–Oppenheimer approximation, i.e., that the electron instantaneously adapts to the ion configuration. It is the rationale behind the Franck–Condon principle, which is not a trivial condition as shown in the article by Repp et al. [23]. They report molecules adsorbed at surface with electronic level spacing of the order of vibration energies and, as a consequence, spoiling the paradigm. In the case of the DB dimer, the electronic states are separated by 1–2 eV while the vibrations involved have the energy of about 13 meV enabling the application of the Franck–Condon principle.

Theory of the electron tunneling through a junction with electron–phonon interactions has been discussed in several papers (e.g., [24, 25]). The latter paper is particularly useful for experimental data interpretation, and it allows simple derivation of the common knowledge that at low temperatures the graph d*I*/d*V* corresponds to the electronic spectral function. The Franck–Condon principle hints at how to calculate this quantity from simple models. The qualitative picture emerging from these calculations is as follows. The incoming electron tunnels to the electronic eigenstate of the neutral geometry. Upon its arrival, there is a sudden change to the potential energy landscape and the relaxation starts. Thus, the electron eigenstate at the neutral geometry is not the eigenstate for the charged system. The ion evolution can be easily written as the ionic ground state of the neutral system expressed in terms of the eigenfunctions of the charged system. The spectral function corresponds to the probability density that a given mode of the charged system will be measured. For the system with large electron-phono coupling it reflects the overlap between ion ground state for the neutral defect and the vibrational modes of the charged system (see Fig. 7). In the harmonic oscillator approximation and for large couplings, this results in the Gaussian shape of the resonance in the STS data centered around the energy value of the electronic state without any electron–phonon coupling. The Gaussian seems to be a generic shape for the large coupling values (see Fig. 7).

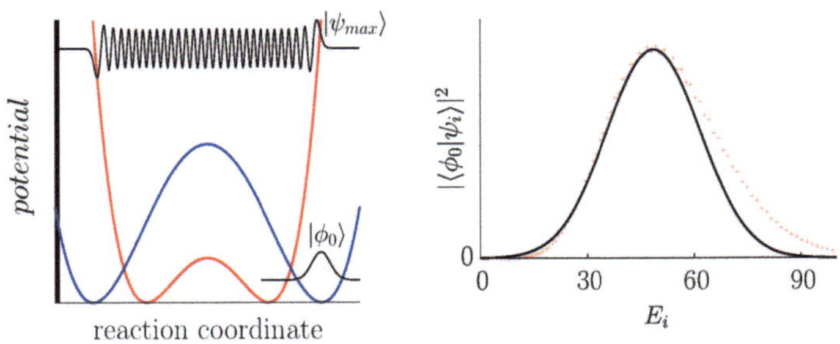

Fig. 7 *Left panel* Schematic one-dimensional potential energy surfaces for the neutral (*blue*) and charged (*red*) defect and the ground state of the neutral defect and highly excited eigenstate of the charged potential with maximal overlap with the ground state. The potentials involve polynomial of the fourth degree and hence are strongly nonlinear. *Right panel* The overlaps between successive eigenmodes of the charged potential and the ground state of the neutral potential. According to the Franck–Condon principle, this quantity corresponds to the spectral function. The points come from numerical calculations, while the solid line corresponds to the fitted Gaussian. It is evident that at lower energies Gaussian fits well to the data and considerable discrepancies are seen for higher energies. STS data shown in Fig. 2 fit into the above scheme

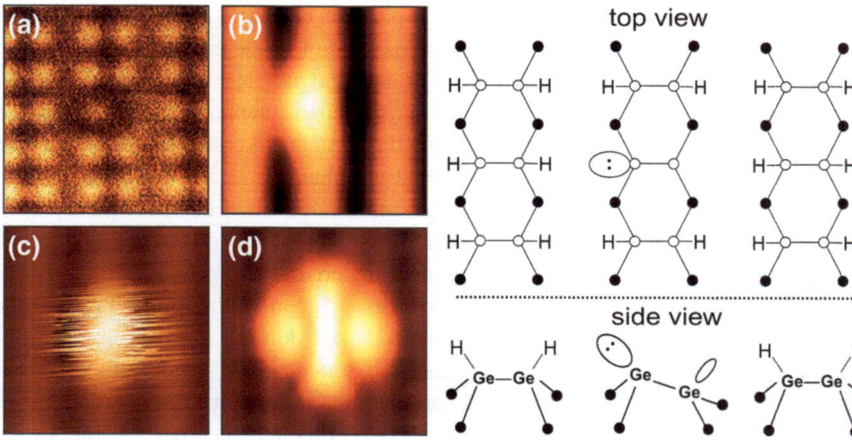

Fig. 8 DB dimer on Ge(001):H surface. *Left panel* **a** constant high image from the noncontact atomic force microscope (*courtesy* b. Such). **b–d** Constant current STM images acquired at 4.5 K for different biases: −0.5 V (**b**), −1.0 V(**c**), +1.3 V (**d**). *Right panel* schematic models of the DB dimer native geometry experimentally imaged in **a** and **b**

AFM and occupied state STM images of a single DB dimer at the Ge(001):H surface are clearly asymmetric (Fig. 8a, b) corroborating the DFT calculations strongly suggesting that its equilibrium geometry is very close to the buckled dimer configuration at the bare surface (Fig. 8 right panel). As the neighboring dimers are symmetric, there is no favorable way of buckling making the defect bistable and allowing switching between these two geometries. Indeed, single flipping events can be observed with STM if states 1 eV below the surface Fermi level are addressed (Fig. 8c). Also, when exploring the unoccupied states such flipping events are induced.

It was the misunderstanding of the empty-state image of DB dimer (Fig. 8d) that prompted interest in the coupling between the unoccupied electron state and the geometric configuration of DB dimer. The problem was that the defect appears bright, large and symmetric. The symmetric images are smooth (Fig. 8c). To arrive at the suggestion that the symmetry originates from flipping events falling out of the apparatus time resolution, it was instrumental to cross-check the empty-state images against the stable occupied state ones as a reference. This revealed that behind the smooth images the geometry changes indeed. These experiments shed light on the previous observation that the position of the unoccupied state rapidly changes with the buckling [2] when calculating the equilibrium geometry of the charged defect. It motivated simulations of the negatively charged defect. It was found that the buckling is substantially reduced for the charged defect and the barrier between the two configuration minima decreased from 0.3–0.4 to about 0.1–0.2 eV. The energy transfer to the ionic degrees of freedom is sufficient to facilitate the flip.

Further evidence for the strong electron–phonon coupling comes from the STS. The peak in the d*I*/d*V* graph attributed to the defect state has an unusual Gaussian shape complying with the strong electron–phonon coupling as described above. Indeed, the interpretation of the vibrational excitation in the case of the DB dimer follows the Franck–Condon paradigm. The last but not least evidence for unresolvable flipping events is the appearance of the DB dimer in STM. As shown in [7], these images can be explained within the simplest Tersoff–Hamann scheme under the assumption that the current virtually makes the dimer flip. Actually, no details as for how the flipping rate depends on the current need to be presumed.

There is a large electron–phonon coupling for the unoccupied state of a single DB dimer. As the number of the dehydrogenated dimers grows, the electron–phonon coupling seems to be of lesser importance. It can be due to the fact that the electronic states form a band-like structure and an incoming electron quickly diffuses. It is possible to acquire stable images of unoccupied states with the characteristic zigzag pattern reflecting the alternative buckling. But, for larger currents, the electron–phonon interactions appear relevant: Frequent geometric changes are observed.

8 Conclusions and Perspectives

In summary, this chapter presents the full protocol of formation of atomic wires on Ge(001):H surface. We start from the detailed description of UHV preparation procedure of hydrogenated Ge(001):H surfaces. The atomically perfect areas of surface act as platforms for atomically precise hydrogen desorption induced by STM tip. Two proposed methods allow effective creation of long atomic chains of DB dimers. Unlike on Si(001):H, the DB states on Ge(001):H can be characterized by STS methods on an undoped substrate. We discuss experimental results showing that the DB electronic states are introduced in the Ge(001):H gap differently for DB lines running perpendicular and parallel to the surface reconstruction rows. DB lines parallel to the surface reconstruction rows display a stronger inter-DB dimer electronic coupling, resulting in a dispersive conduction band spanning 0.7 eV for an infinite parallel DB line. The electronic behavior of short DB lines in both directions is of great importance for the design of more complex DB structures which could act as independent logic circuits or atomic-scale interconnects. Moreover, discussed strong electron–phonon coupling effect, which is observed on single DB dimer, must be taken into account in a design of the future devices as well.

The next step toward realization of DB-based atomic-scale interconnects vision is the realization of planar transport measurements on that systems. This goal could be achieved with the use of the new ScientaOmicron 4-probe LT-STM [8]. Figure 9 shows a long DB wire already prepared with the use of this machine by application of hydrogen desorption presented in this chapter. This result proves feasibility of such a direct multiprobe STM characterization of atomic wires on hydrogenated semiconductor surfaces and opens a new field of exiting experimental and theoretical research.

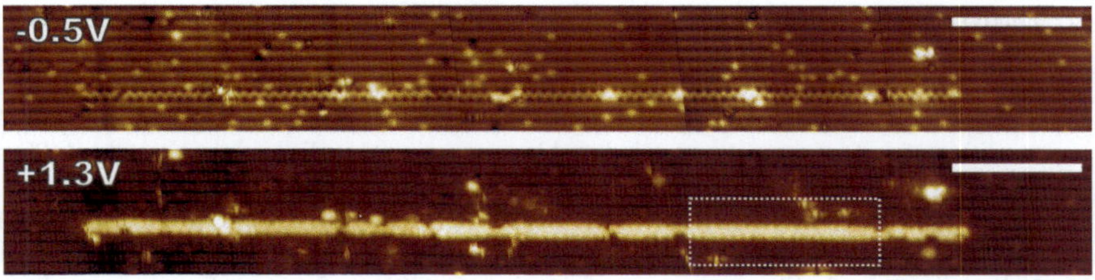

Fig. 9 DB dimer wire on Ge(001):H surface formed and imaged by one of the 4-probe STM scanners at 4.5 K. The ~70-nm-long wire consists of 156 bare Ge dimers (DB dimers) and has 14 atomic-scale defects including 9 single Ge atoms (single DBs) and 5 unknown defects (adsorbates or vacancies). The structure is constructed by several line desorption procedures (STM feedback loop closed). For both images, tunneling current is set to 50 pA and the scale bar length is 10 nm. *Dotted rectangle* shows perfect DB dimer wire part of ~15 nm length

Acknowledgements We acknowledge European Union Collaborative ICT Projects: "Atomic Scale and Single Molecule Logic Gate Technologies" (ATMOL, contract no. 270028) and "Planar Atomic and Molecular Scale devices" (PAMS, contract no. 610446). MK acknowledges financial support received from the Foundation for Polish Science (FNP).

References

1. Prauzner-Bechcicki, J.S., Godlewski, S., Szymonski, M.: Atomic- and molecular-scale devices and systems for single-molecule electronics. Physica Status Solidi (a) **209**(4), 603–613 (2012)
2. Kolmer, M., et al.: Electronic properties of STM-constructed dangling-bond dimer lines on a Ge(001)-(2x1):H surface. Phys. Rev. B **86**(12), 125307 (2012)
3. Kolmer, M., et al.: Atomic scale fabrication of dangling bond structures on hydrogen passivated Si(001) wafers processed and nanopackaged in a clean room environment. Appl. Surf. Sci. **288**, 83–89 (2014)
4. Kolmer, M., et al.: Realization of a quantum Hamiltonian Boolean logic gate on the Si(001): H surface. Nanoscale **7**(29), 12325–12330 (2015)
5. Kepenekian, M., et al.: Surface-state engineering for interconnects on H-passivated Si(100). Nano Lett. **13**(3), 1192–1195 (2013)
6. Godlewski, S., et al.: Dynamical behavior of a dangling bond dimer on a hydrogenated semiconductor: Ge(001):H. Phys. Rev. B **92**(11), 115403 (2015)
7. Engelund, M., et al.: The butterfly—a well-defined constant-current topography pattern on Si(001):H and Ge(001):H resulting From current-induced defect fluctuation. Phys. Chem. Chem. Phys. **18**, 19309–19017 (2016)
8. Yang, J.S., et al.: Imaging, single atom contact and single atom manipulations at low temperature using the new ScientaOmicron LT-UHV-4 STM. Eur. Phys. J. Appl. Phys. **73**(1), 10702 (2016)
9. Hummer, K., Harl, J., Kresse, G.: Heyd-Scuseria-Ernzerhof hybrid functional for calculating the lattice dynamics of semiconductors. Phys. Rev. B **80**(11), 115205 (2009)
10. Radny, M.W., et al.: Valence surface electronic states on Ge(001): reply. Phys. Rev. Lett. **103** (18), 189702 (2009)
11. Kamiyama, E., Sueoka, K., Vanhellemont, J.: Surface-induced charge at a Ge (100) dimer surface and its interaction with vacancies and self-interstitials. J. Appl. Phys. **113**(9), 093503 (2013)

12. Zandvliet, H.J.W.: The Ge(001) surface. Physics Reports-Review Section of Physics Letters **388**(1), 1–40 (2003)
13. Mönch, W.: Semiconductor Surfaces and Interfaces. Springer, Berlin (2001)
14. Takagi, Y., et al.: Rewritable nanopattern on a Ge(001) surface utilizing p(2x2)-to-c(4x2) transition of surface reconstruction induced by a scanning tunneling microscope. Appl. Phys. Lett. **84**(11), 1925–1927 (2004)
15. Seo, H.S., et al.: Critical differences in the surface electronic structure of Ge(001) and Si(001): Ab initio theory and angle-resolved photoemission spectroscopy. Phys. Rev. B **89**(11) (2014)
16. Nakatsuji, K., et al.: Electronic states of the clean Ge(001) surface near Fermi energy. Phys. Rev. B **72**(24), 241308 (2005)
17. Sagisaka, K., Fujita, D.: Standing waves on Si(100) and Ge(100) surfaces observed by scanning tunneling microscopy. Phys. Rev. B **72**(23), 235327 (2005)
18. Landemark, E., et al.: Electronic-structure of clean and hydrogen-chemisorbed Ge(001) surfaces studied by photoelectron-spectroscopy. Phys. Rev. B **49**(23), 16523–16533 (1994)
19. Jeon, C., et al.: Evidence from ARPES that the Ge(001) surface is semiconducting at room temperature. Phys. Rev. B **74**(12), 125407 (2006)
20. Zandvliet, H.J.W., Vansilfhout, A., Sparnaay, M.J.: Metallic properties of the Ge(001) surface. Phys. Rev. B **39**(8), 5576–5578 (1989)
21. Wojtaszek, M., et al.: Fermi level pinning at the Ge(001) surface—a case for non-standard explanation. J. Appl. Phys. **118**(18), 185703 (2015)
22. Wojtaszek, M., et al.: Inversion layer on the Ge(001) surface from the four-probe conductance measurements. Appl. Phys. Lett. **105**(4), 042111 (2014)
23. Repp, J., Liljeroth, P., Meyer, G.: Coherent electron-nuclear coupling in oligothiophene molecular wires. Nat. Phys. **6**(12), 975–979 (2010)
24. Wingreen, N.S., Jacobsen, K.W., Wilkins, J.W.: Inelastic-scattering in resonant tunneling. Phys. Rev. B **40**(17), 11834–11850 (1989)
25. Jonson, M.: Quantum-mechanical resonant tunneling in the presence of a Boson field. Phys. Rev. B **39**(9), 5924–5933 (1989)

Si(100):H and Ge(100):H Dimer Rows Contrast Inversion in Low-temperature Scanning Tunneling Microscope Images

Hiroyo Kawai, Tiong Leh Yap, Olga Neucheva, Marek Kolmer, Marek Szymoński, Cedric Troadec, Mark Saeys and Christian Joachim

Abstract Detailed low-temperature scanning tunneling microscope images of the Si(100)-2×1-H and the Ge(100)-2×1-H surfaces show a remarkable contrast inversion between filled- and empty-state images where the hydrogen dimer rows appear bright for filled-state images and dark for empty-state images. This contrast

---

Sections 1, 3, and 5: Reprinted with minor editorial changes from Surface Science Letters, **632**, Tiong Leh Yap, Hiroyo Kawai, Olga Neucheva, Andrew Thye Shen Wee, Cedric Troadec, Mark Saeys, and Christian Joachim, "Si(100)-2×1-H dimer rows contrast inversion in low-temperature scanning tunneling microscope images," L13-L17, 2015, with permission from Elsevier.

H. Kawai · T.L. Yap · O. Neucheva · C. Troadec
Institute of Materials Research and Engineering, 2 Fusionopolis Way, Innovis, #08-03, Singapore 138634, Singapore

T.L. Yap
GLOBALFOUNDRIES Singapore Pte Ltd., 60 Woodlands Industrial Park D Street 2, Singapore 738406, Singapore

T.L. Yap
Department of Physics, National University of Singapore, 2 Science Drive 3, Singapore 117542, Singapore

M. Kolmer · M. Szymoński
Faculty of Physics, Astronomy and Applied Computer Science, Center for Nanometer-Scale Science and Advanced Materials, NANOSAM, Jagiellonian University, Lojasiewicza 11, PL 30-348 Krakow, Poland

M. Saeys
Laboratory for Chemical Technology, Ghent University, Technologie Park 914, 9052 Ghent, Belgium

C. Joachim (✉)
GNS & MANA Satellite, CEMES-CNRS, 29 Rue J. Marvig, 31055 Toulouse Cedex, France
e-mail: joachim@cemes.fr

C. Joachim
International Center for Materials Nanoarchitectronics (WPI-MANA), National Institute for Materials Science (NIMS), 1-1 Namiki, Tsukuba, Ibaraki 305-0044, Japan

M. Kolmer and C. Joachim (eds.), *On-Surface Atomic Wires and Logic Gates*, Advances in Atom and Single Molecule Machines,
DOI 10.1007/978-3-319-51847-3_4

inversion originates from the change in the dominant surface states and their coupling to the tip apex and the bulk channels as a function of the bias voltage.

1 Introduction

Hydrogen-passivated semiconductor surfaces such as Si(100):H and Ge(100):H are an important platform for the construction of atomic-scale circuits [1, 2] and molecule-based devices [3] since they are robust surfaces [4] with a low defect density [5]. In addition, Si(100):H has a relatively large surface band gap of 2.1 eV [6]. The Si(100):H and Ge(100):H surfaces have therefore been studied extensively by low-temperature ultrahigh vacuum scanning tunneling microscopy (LT-STM) [7–10]. Surface hydrogen atoms of the Si–Si or Ge–Ge dimer rows can be desorbed with atomic precision by pulsing the STM tip bias voltage to create surface dangling bonds. The surface dangling bonds introduce well-defined states in the Si (100):H or Ge(100):H surface band gap and have been proposed as building blocks for single atom transistors [2], quantum dots [11], quantum wells [12], and logic gates [13, 14].The atom-by-atom construction of dangling-bond devices requires the precise location of the surface hydrogen atoms. As demonstrated in this chapter, the STM contrast of the Si(100):H and the Ge(100):H surface depends on the imaging conditions (bias voltage and tip structure) and does not always match the atomic-scale structure of the surface. At positive bias voltage, the inter-dimer valleys appear brighter than the topologically higher dimer rows. This contrast inversion leads to an apparent half row shift of the surface atomic-scale structure and highlights that the determination of the atomic-scale structure of atomic wires and circuits constructed on the Si(100):H or Ge(100):H surface using STM images needs to be done with care.

In this chapter, we demonstrate the bias voltage and tip-dependent contrast inversion of the Si(100):H surface. By comparing the STM experimental images with the calculated images, we show that this contrast inversion originates from the different dominant surface states and their coupling to the tip apex and the bulk silicon channels as a function of the bias voltage. The contrast inversion of the Ge(100):H surface is also discussed briefly in the subsequent section.

2 Experimental and Calculation Details

To analyze and elucidate the bias voltage and tip-dependent contrast inversion of the Si(100):H and Ge(100):H surfaces, LT-UHV STM images of a hydrogen-passivated surfaces were recorded at liquid helium temperature (4.5 K) using a ultrahigh vacuum low-temperature STM machines (Omicron Nanotechnology GmbH) with a base pressure of 2×10^{-10} mbar [15]. Si(100) samples were obtained from Sb-doped n-type wafers and B-doped p-type wafers both with a resistivity <0.1 Ω cm. The Si(100):H surface was prepared by annealing at 600 °C for 12 h and several cycles of flashing to 1200 °C for 10 s. After this, the substrate temperature was reduced to 375 °C and atomic hydrogen was dosed up to 13.4 L (1 L = 1 × 10^{-6} Torrs) at a H_2 partial pressure of 2 × 10^{-8} mbar. The H_2 gas (>99.95% purity) was introduced via a leak valve and cracked by passing over a tungsten filament at 35 mA, about 15 cm above the sample. The preparation of the Ge(001) surface (0.5 mm thick, undoped wafers) consisted of cycles of 1 keV Ar^+ sputtering for 15 min with the sample kept at 760 °C. The hydrogenation procedure was performed for 10 min with the sample kept at 200 °C and the hydrogen partial pressure maintained at 1 × 10^{-7} mbar level.

STM images of both surfaces were computed using the surface Green-function matching (SGFM) method [16] with an extended Hückel molecular orbital (EHMO) Hamiltonian [17] tuned to match HSE06 density functional theory band structures [18]. The STM junction was modeled as a semi-infinite W(111) slab, a Si-terminated STM tip, a five-layer Si(100):H surface, and the semi-infinite Si(100) bulk, as illustrated in Fig. 3a. In the case of Ge(100):H surface, a Ge-terminated STM tip, a nine-layer Ge(001):H surface and the semi-infinite Ge(001) bulk was used.

3 Contrast Inversion of the Si(100):H Surface

Figure 1a–d shows STM images of n-type Si(100):H with a setpoint current of 20 pA and for a large range of bias voltages. The corresponding line-scan profiles perpendicular to the dimer rows are shown in Fig. 1e. The filled-state image (Fig. 1d) shows the 2 × 1 dimer rows with a maximum at the center of the Si dimer rows and a minimum at the inter-row valleys (Fig. 1d, e), matching the surface structure. The filled-state images are relatively insensitive to the change in bias voltage. A more interesting variation of the contrast with the bias voltage is observed for the empty-state images. At a low-positive bias voltage (+1.0 V, Fig. 1c), the contrast is similar to the filled-state contrast and matches the surface geometry, except for the presence of a shallow local minimum instead of a maximum at the center of the dimer rows. When the bias voltage is increased, the local minimum at the inter-row valleys becomes shallower and at +1.3 V, a fully symmetric 1 × 1 structure with equally spaced peaks is observed (Fig. 1b, e). For a bias voltage above +1.3 V, the 2 × 1 pattern reemerges; however, the minimum is now at the center of the dimer rows, and maximum is found at the center of inter-dimer valleys, a complete inversion of the contrast (Fig. 1a). The contrast inversion is more clearly shown in Fig. 2, recorded for p-type Si(100):H, where the bias was switched suddenly between −1.6 and +1.6 V while scanning. The reversibility of the contrast inversion demonstrates that piezo creep or thermal drift does not cause the contrast inversion. Experiments performed with both n- and p-type Si(100):H demonstrate that the inversion is independent of the doping.

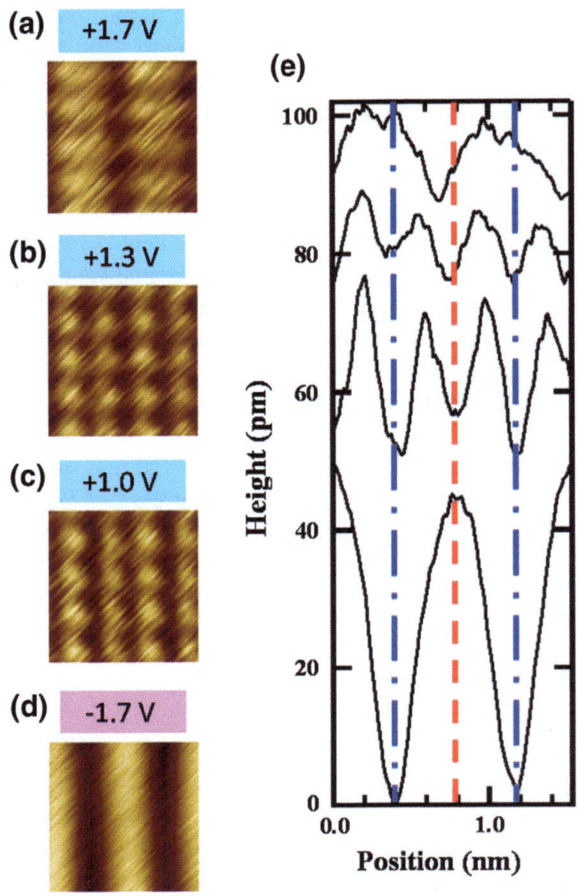

Fig. 1 **a–d** STM images of the n-type Si(100)-2×1-H surface for different bias voltages (20 pA current, 1.54 nm × 1.54 nm scan area). **e** Corrugation profiles perpendicular to dimer rows for the images in (**a**)–(**d**) from *top* to *bottom*. The minima and maxima in the corrugation are indicated to illustrate the contrast inversion

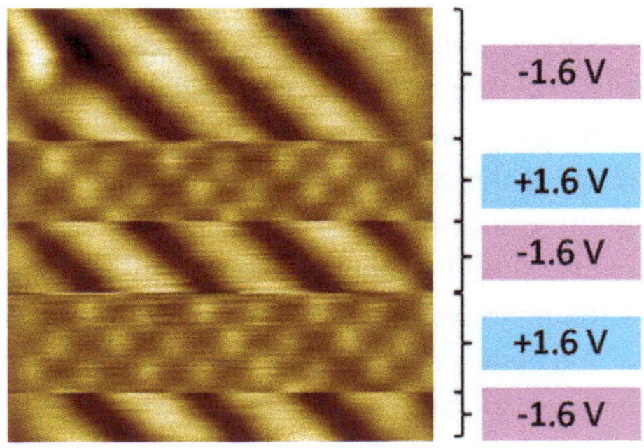

Fig. 2 STM image of p-type Si(100)-2×1-H. The bias voltage was suddenly switched from −1.6 V (filled state) to +1.6 V (empty state) while scanning. (20 pA current, 4 nm × 4 nm scan area)

The d*I*/d*V* spectra for both the n- and the p-type Si(100):H surface do not show signature peaks for surface dangling bond or defect states (not shown), in agreement with previous spectra [6, 19], suggesting that such states do not contribute to the contrast inversion. The effect of tip apex-to-sample distance on the image was explored as well. The contrast inversion in the empty-state images was also observed for a bias voltage of +1.8 V and setpoint currents between 400 and 500 pA, showing that the contrast inversion is observed for a small range of tip-sample distances.

To understand the origin of the observed contrast inversion, constant current images and *T*(*E*) electronic transmission spectra through the tunnel junction were computed for the STM junction shown in Fig. 3a. The calculated *T*(*E*) spectra are shown for two tip apex positions: at the center of the Si dimer rows (Fig. 3b, on dimer, dotted red curve) and at the center of the inter-dimer valleys (Fig. 3b, between dimers, blue curve). In the valence band region (filled states) and at the center of the dimers, *T*(*E*) is always higher than *T*(*E*) at the center of the valley, resulting in images which are bright above the dimer rows and darker above the valleys (Fig. 4e), in agreement with the experimental images. Although the calculated corrugation varies with the bias voltage, the positions of the maxima and minima in the calculated filled-state images do not change with bias. In the conduction band region (empty states), the contrast depends strongly on the bias voltage, as seen from the crossings of the transmission spectra for the two tip positions.

At the conduction band edge, *T*(*E*) at the center of the dimers is slightly lower than *T*(*E*) at the center of the valleys, leading to a contrast inversion in the calculated constant current images for a bias voltage of +0.80 V (Fig. 4d). At +1.0 eV, *T*(*E*) above the dimer rows and *T*(*E*) above the center of the valleys are comparable, and above +1.1 eV, *T*(*E*) above the dimer rows becomes higher, except for a resonance peak at 2.8 eV. As a result, the calculated contrast reverts for bias voltages above +1.2 V (Fig. 4c) and again matches the surface structure for a bias voltage of +1.7 V (Fig. 4b). At a higher bias of +3.0 V, a second contrast inversion, which is caused by contributions from the resonance at 2.9 eV, is observed (Fig. 4a). The calculations show that the empty-state images result from a competition between states localized at the dimers and states localized at the valleys. Their coupling to the STM tip and to the Si bulk states determines the contrast in the calculated images. The line-scan profiles perpendicular to the dimer rows (Fig. 4f) more clearly show the various transitions in the calculated contrast.

The band structure of the Si(100):H surface and the LDOS near the conduction and valence band edges were calculated for a 13-layer slab to identify the states which contribute to the empty- and filled-state images (Fig. 5a–g). The LDOS at valence band edge (Fig. 5b, −0.74 eV) shows bonding Si–Si surface states along the dimer rows, which are well coupled to the bulk states as shown by their expansion into the bulk layers. These surface states contribute to the appearance of a peak at the center of the dimer rows in the filled-state images. On the other hand, the LDOS at the conduction band edge shows Si–Si anti-bonding states in the bulk. Near the surface, these states greatly enhance the intensity of the LDOS in the valley between the dimer rows (Fig. 5d, e). The coupling of these states to the STM tip contributes to the contrast inversion observed in the calculated images below 1.2 V. At higher energies, Si–H anti-bonding states are found (Fig. 5g). These states are likely responsible for the images calculated near +1.2 V (Fig. 4c).

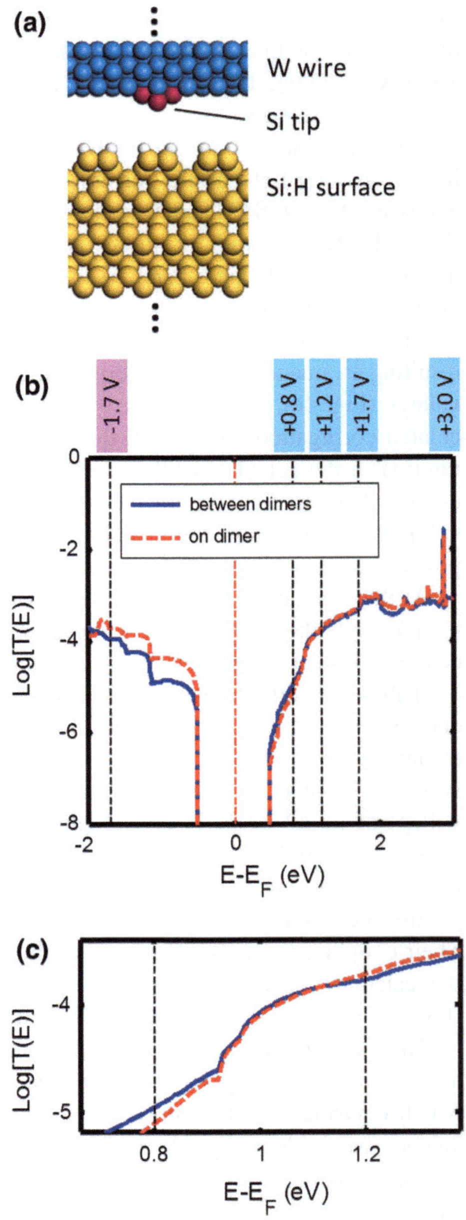

Fig. 3 **a** Tunneling junction used to model the STM images and the transmission coefficient, *T*(*E*). **b** *T*(*E*) calculated for a tip above the center of the dimers (*dotted red curve*) and above the center of the inter-dimer valleys (*blue curve*) for a tip height of 6 Å. **c** Zoom-in for *T*(*E*) near the conduction band edge, showing the crossings between *T*(*E*) above the center of the dimers (*dotted red curve*) and *T*(*E*) above the center of the inter-dimer valleys (*blue curve*)

4 Contrast Inversion for the Ge(100):H Surface

A similar contrast inversion has been observed for the Ge(100):H surface. The STM images for the Ge(100):H surface, obtained at bias voltages of −0.5, +0.7, +1.2 and +1.4 V, are shown in Fig. 6a–d, respectively. The line-scan profile for each surface

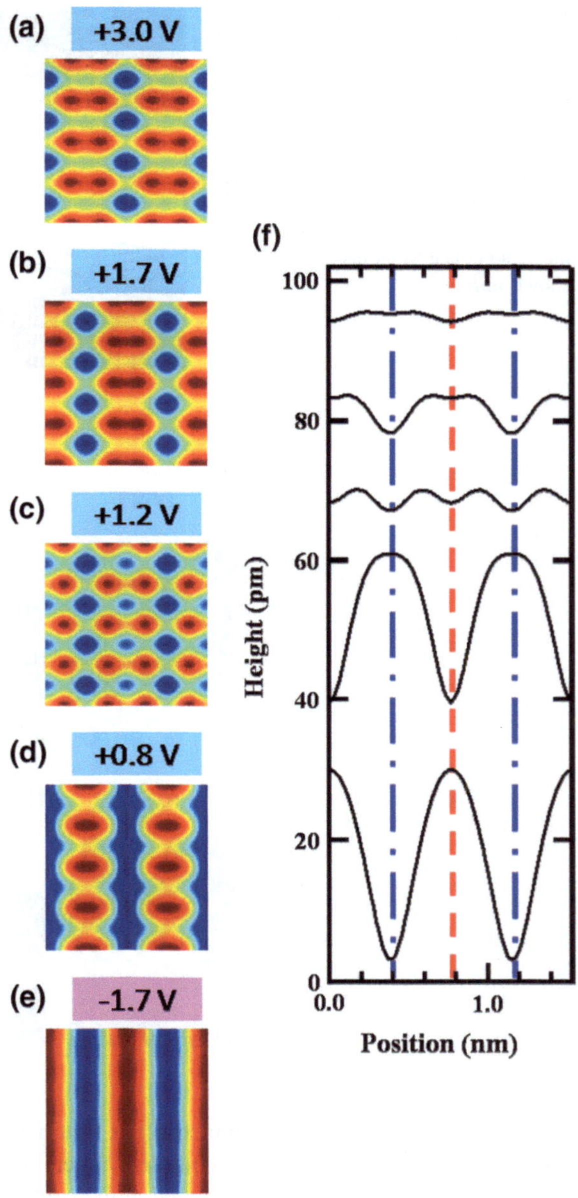

Fig. 4 **a–e** Constant current STM images calculated for a current of 20 pA, scan size of 1.54 nm × 1.54 nm. **f** Line-scan profiles perpendicular to the dimer rows for the images in (**a**)–(**e**) from *top* to *bottom*

taken at the line indicated in the image is shown in Fig. 6e (green, blue, orange, and red lines correspond to the line-scans obtained at −0.5, +0.7, +1.2 and +1.4 V, respectively). The bright spot at top right corner of each image corresponds to a bare Ge dimer, which was used as a position marker. The images at −0.5 and +0.7 V are similar, with maximum height at the center of the dimer row (Fig. 6a, b). At a higher positive bias voltage of +1.2 V, there is a minimum at the center of

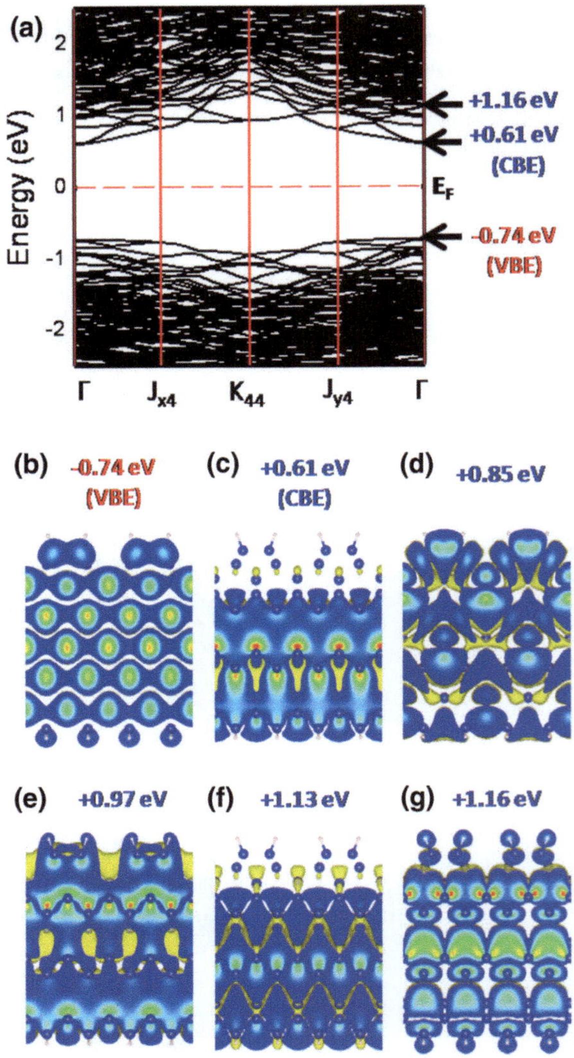

Fig. 5 **a** Band structure for a 13-layer Si(100):H slab using a $p(4 \times 4)$ unit cell. **b–g** Selected LDOS plots at the conduction and valence band edge for the gamma point and for energies indicated in the band structure in (**a**). The colors indicate the intensity of the LDOS. The Si atoms are shown in *blue*

the dimer row, resulting in a symmetric 1×1 structure (Fig. 6c, e). At even higher positive bias voltage (+1.4 V), complete inversion of the contrast is observed, where the maximum height is observed at the center of the inter-dimer valleys and the minimum at the center of the dimer row (Fig. 6d, e). This trend is very similar to the trend observed for the Si(100):H surface, except that the non-inverted image at very small positive bias voltage could not observed for Si(100):H due to the larger band gap for the Si(100):H surface.

The STM images for the Ge(100):H surface were calculated for different bias voltages and compared with the experimental STM images. Figure 7a–c shows the STM images of the Ge(100):H surface calculated at +1.0, +0.5 and −0.5 V,

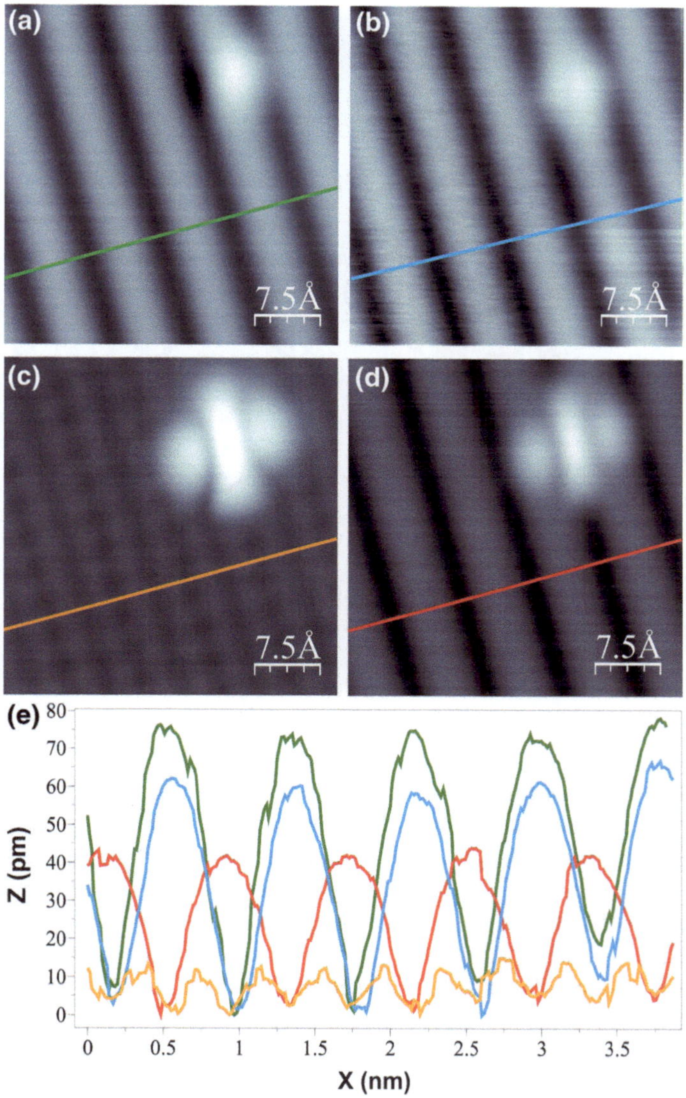

Fig. 6 STM images of the Ge(100):H surface obtained at **a** −0.5 V, **b** +0.7 V, **c** +1.2 V, and **d** +1.4 V. The corresponding line-scan profiles are shown in (**e**). *Green*, *blue*, *orange*, and *red lines* correspond to the line-scans obtained at −0.5 V (*green*), +0.7 V (*blue*), +1.2 V (*yellow*), and +1.4 V (*red*), respectively. The line-scans were taken at the lines indicated in (**a**)–(**d**) for each surface

respectively. The atomic structure of the Ge(100):H surface is shown in Fig. 7d. The line-scan profiles perpendicular to the dimer rows are shown in Fig. 7e from top to bottom. At a negative bias voltage of −0.5 V, the maximum height is found at the center of the dimer row (Fig. 7a, e), and similar images were obtained at more

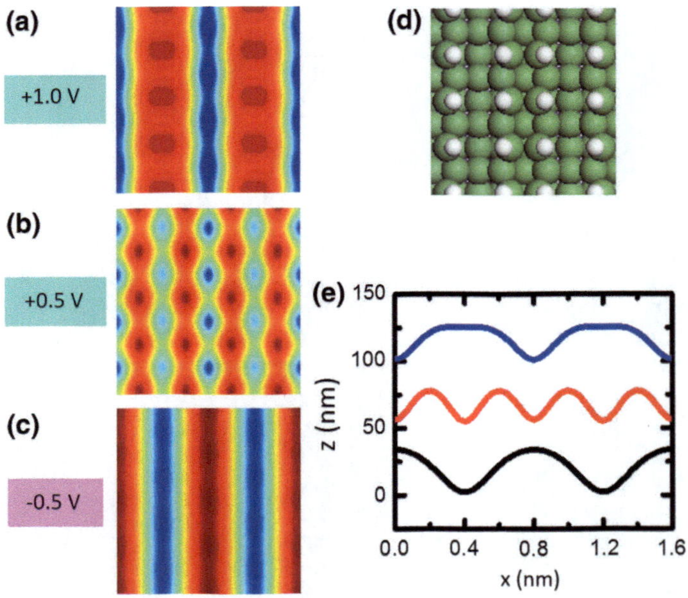

Fig. 7 Calculated STM images of the Ge(100):H surface at **a** +1.0 V, **b** +0.5 V and **c** −0.5 V. **d** Atomic structure of the Ge(100):H surface used in the calculation. **e** Line-scan profiles perpendicular to the dimer rows for the images in **a–c**, from +1.0 V (*blue*), +0.5 V (*red*), −0.5 V (*black*)

negative bias voltages (not shown). At a small positive bias of +0.5 V, the symmetric 1×1 structure is obtained (Fig. 7b, e), while at a more positive bias of +1.0 V, the maximum height shifts to the inter-dimer valleys, leading to an inversion of the contrast (Fig. 7c, e). In agreement with the experimental images, the calculated STM image contrast is inverted for higher positive biases. Unfortunately, the calculations did not reproduce the non-inverted contrast for small positive biases. This might be related to tip-induced band bending which was not included in the calculations. Tip-induced band bending shifts the surface states at conduction band edge that lead to the 1×1 structure image to slightly higher energies. As a consequence, the image at a smaller positive bias voltage would resemble the non-inverted image observed at negative bias voltages.

5 Conclusions

We have shown that the experimentally observed contrast inversion in the empty-state LT-STM images of the robust Si(100):H surface results from a competition between surface states spatially located at the dimer rows and states spatially located above the inter-dimer valleys. The contrast depends on the bias voltage as well as the tip character. The contrast observed for the filled-state images matches the geometry of the dimer rows and is insensitive to the bias voltage and the tip character. For certain bias conditions, empty-state images allow to precisely image the hydrogen atoms; however, filled-state images provide the reference to determine the exact location of the dimer rows on the Si(100)-2×1-H surface, a very important determination for the construction of dangling bonds on the Si(100): H surface. A similar trend was observed for the Ge(100):H surface. Thus, care is required when constructing and interpreting the atomic structure of dangling-bond structures on the Si(100):H and the Ge(100):H surfaces.

Acknowledgements We acknowledge the Agency of Science, Technology, and Research (A*STAR) for funding provided through the Visiting Investigatorship Programme Atom Technology project 1021100972, and through the AtMol integrated project contract number 270028 from the European Commission. We also acknowledge the A*STAR Computational Resource Centre (A*CRC) for computational resources and support. MK acknowledges financial support received from the Foundation for Polish Science (FNP).

References

1. Joachim, C., Martrou, D., Rezeq, M., Troadec, C., Deng, J., Chandrasekhar, N., Gauthier, S.: Multiple atomic scale solid surface interconnects for atom circuits and molecule logic gates. J. Phys. Condens. Matter **22**, 084025 (2010). doi:10.1088/0953-8984/22/8/084025
2. Fuechsle, M., Miwa, J.A., Mahapatra, S., Ryu, H., Lee, S., Warschkow, O., Hollenberg, L.C. L., Klimeck, G., Simmons, M.Y.: A single-atom transistor. Nat. Nanotechnol. **7**, 242 (2012). doi:10.1038/nnano.2012.21
3. Piva, P.G., DiLabio, G.A., Pitters, J.L., Zikovsky, J., Rezeq, M., Dogel, S., Hofer, W.A., Wolkow, R.A.: Field regulation of single-molecule conductivity by a charged surface atom. Nature **435**, 658 (2005). doi:10.1038/nature03563
4. Hersam, M.C., Guisinger, N.P., Lyding, J.W., Thompson, D.S., Moore, J.S.: Atomic-level study of the robustness of the Si(100)-2×1: H surface following exposure to ambient conditions. App. Phys. Lett. **78**, 886 (2001). doi:10.1063/1.1348322
5. Boland, J.J.: Scanning tunneling microscopy of the interaction of hydrogen with silicon surfaces. Adv. Phys. **42**, 129 (1993). doi:10.1080/00018739300101474
6. Bellec, A., Riedel, D., Dujardin, G., Boudrioua, O., Chaput, L., Stauffer, L., Sonnet, P.: Electronic properties of the n-doped hydrogenated silicon (100) surface and dehydrogenated structures at 5 K. Phys. Rev. B **80**, 245434 (2009). doi:10.1103/PhysRevB.80.245434
7. Buehler, E.J., Boland, J.J.: Dimer preparation that mimics the transition state for the adsorption of H_2 on the Si(100)-2×1 surface. Science **290**, 506 (2000). doi:10.1126/science.290.5491.506
8. Bellec, A., Riedel, D., Dujardin, G., Rompotis, N., Kantorovich, L.N.: Dihydride dimer structures on the Si(100): H surface studied by low-temperature scanning tunneling microscopy. Phys. Rev. B **78**, 165032 (2008). doi:10.1103/PhysRevB.78.165302
9. Labidi, H., Kantorovich, L., Riedel, D.: Atomic-scale control of hydrogen bonding on a bare Si(100)-2×1 surface. Phys. Rev. B **86**, 165441 (2012). doi:10.1103/PhysRevB.86.165441
10. Kolmer, M., Godlewski, S., Kawai, H., Such, B., Krok, F., Saeys, M., Joachim, C., Szymonski, M.: Electronic properties of STM-constructed dangling-bond dimer lines on a Ge(001)-(2×1):H surface. Phys. Rev. B. **86**, 215307 (2012). doi:10.1103/PhysRevB.86.125307

11. Haider, M.B., Pitters, J.L., DiLabio, G.A., Livadaru, L., Mutus, J.Y., Wolkow, R.A.: Controlled coupling and occupation of silicon atomic quantum dots at room temperature. Phys. Rev. Lett. **102**, 046805 (2009). doi:10.1103/PhysRevLett.102.046805
12. Schofield, S.R., Studer, P., Hirjibehedin, C.F., Curson, N.J., Aeppli, G., Bowler, D.R.: Quantum engineering at the silicon surface using dangling bonds. Nat. Commun. **4**, 1649 (2013). doi:10.1038/ncomms2679
13. Kawai, H., Ample, F., Wang, Q., Yeo, Y.K., Saeys, M., Joachim, C.: Dangling-bond logic gates on a Si(100)-(2×1)-H surface. J. Phys. Condens. Matter **24**, 095011 (2012). doi:10.1088/0953-8984/24/9/095011
14. Ample, F., Duchemin, I., Hliwa, M., Joachim, C.: Single OR molecule and OR atomic circuit logic gates interconnected on a Si(100)H surface. J. Phys. Condens. Matter **23**, 125303 (2011). doi:10.1088/0953-8984/23/12/125303
15. Neucheva, O.A., Thamankar, R.M., Yap, T.L., Troadec, C., Deng, J., Joachim, C.: Atomic scale interconnection machine. In: Joachim, C., (ed.) Atomic Scale Interconnection Machines, Advances in Atom and Single Molecule Machines, pp. 23–33. Springer, Heidelberg. doi:10.1007/978-3-642-28172-3_3 (2012)
16. Cerda, J., Hove, M.A.V., Sautet, P., Salmeron, M.: Efficient method for the simulation of STM images. I. Generalized green-function formalism. Phys. Rev. B **56**, 15885 (1997). doi:10.1103/PhysRevB.56.15885
17. Kienle, D., Bevan, K.H., Liang, G.-C., Siddiqui, L., Cerda, J.I., Ghosh, A.W.: Extended Hückel Theory for band structure, chemistry, and transport II. Silicon. J. Appl. Phys. **100**, 043715 (2006). doi:10.1063/1.2259820
18. Moussa, J.E., Schultz, P.A., Chelikowsky, J.R.: Analysis of the Heyd-Scuseria-Ernzerhof density functional parameter space. J. Chem. Phys. **136**, 204117 (2012). doi:10.1063/1.4722993
19. Kolmer, M., Godlewski, S., Zuzak, R., Wojtaszek, M., Rauer, C., Thuaire, A., Hartmann, J.-M., Moriceau, H., Joachim, C., Szymonski, M.: Atomic scale fabrication of dangling bond structures on hydrogen passivated Si(001) wafers processed and nanopackaged in a clean room environment. Appl. Surf. Sci. **288**, 83 (2014). doi:10.1016/j.apsusc.2013.09.124

Band Engineering of Dangling-Bond Wires on the Si(100)H Surface

Roberto Robles, Michael Kepenekian, Christian Joachim, Ricardo Rurali and Nicolas Lorente

Abstract Nanoscale devices need to be connected among them and to the macroscopic world. Ideally, interconnects embedded in the environment holding the device will have an undeniable practical and technological advantage. Since silicon surfaces dominate nano- and micro-technologies, the crafting of nanowires on these surfaces has been suggested to be the way to create effective interconnecting wires. Here, we review the work done on dangling-bond wires formed by removing passivated agents from the Si(100)-H surface. These wires are formed by adjacent dangling bonds that rehybridize and create a surface-confined band structure capable of driving charge and spin along the surface. Unfortunately, the 1-D character of these wires leads to instabilities that create band gaps. The way to go is then to engineer the band gaps so as to create low-gap systems or directly

Reprinted (adapted) with permission from M. Kepenekian, R. Robles, C. Joachim and N. Lorente, Nano Lett., 13, 1192 (2013), DOI:10.1021/nl304611m, Copyright 2013 American Chemical Society. Figures 1, 2 and 3 are reproduced from Mikaël Kepenekian et al. Nanotechnology, 25, 465703 (2014) with permission granted by IOP Publishing London on the 18 Novembre 2016.

R. Robles
Catalan Institute of Nanoscience and Nanotechnology (ICN2), CSIC and the Barcelona Institute of Science and Technology, Campus UAB Bellaterra, 08193 Barcelona, Spain

M. Kepenekian
Institut Des Sciences Chimiques de Rennes, UMR 6226CNRS—Université de Rennes 1—Ecole Nationale Supérieure de Chimie de Rennes, Rennes, France

C. Joachim
Centre D'Elaboration Des Materiaux et D'Etudes Structurales (CEMES) CNRS, 29 Rue J. Marvig, 31055 Toulouse, Cédex, France

R. Rurali
Institut de Ciència de Materials de Barcelona CSIC, 08193 Bellaterra, Spain

N. Lorente (✉)
Centro de Física de Materiales CFM/MPC (CSIC-UPV/EHU),
Paseo Manuel de Lardizabal 5 and Donostia International Physics Center (DIPC),
Paseo Manuel de Lardizabal 4, 20018 Donostia-San Sebastián, Spain
e-mail: nicolas.lorente@ehu.eus

M. Kolmer and C. Joachim (eds.), *On-Surface Atomic Wires and Logic Gates*,
Advances in Atom and Single Molecule Machines,
DOI 10.1007/978-3-319-51847-3_5

metallic wires. Two strategies are reviewed in this chapter. One is the atom manipulation of wires to find the geometries that favored low-gap dangling-bond wires. The second one is the use of doping to change the electronic properties of the wires.

1 Introduction

Bottom-up approaches are revolutionary because of their enhanced control over nanostructuring [1]. Interconnects between active atomic parts of a device have been created using mechanical, [2] electromigration [3] and scanning-probe techniques [4–6]. To this respect, the use of silicon as the substrate that holds device and interconnects is very interesting due to its properties [7–12]. Nanowires are of great use for the creation of technology on the atomic scale. Silicon nanowires have seen a lot of activity in the last years [13, 14]. An alternative approach to 3-D nanowires is the crafting of nanowires on a holding Si substrate [15–17]. Here, the working principle is that the available electronic states are localized to a few dangling bonds on the surface preventing any contact with bulk states and permitting the hopping of electrons from one dangling bond (DB) to the next one [16–23].

Despite these promising features, the 1-D nature of DB wires leads to strong instabilities that create electronic gaps in the wire's electronic structure. Indeed, any perturbation destroys the metalicity of the state opening a band gap. For example, the electron-vibration coupling leads to a Peierls distortion with visible consequences in the arrangement of atoms. Electron–electron correlations are also present, possibly leading to a Luttinger liquid phase, and to magnetic correlations [24]. Experimental data have been gathered for finite wires [25–27]. The visualization of deformed structures with the STM [25, 27] seems to indicate the existence of strong electron-vibration couplings and the appearance of a Jahn-Teller effect (the counter-part of the extended-system Peierls effect). If an extra charge is included to account for transport, these effects are enhanced and the injected electron propagates with the wire deformation leading to small polaron transport [28, 29]. Moreover, soliton transport has also been predicted as due to the displacement of the domain wall between adjacent domains of distorted atoms [29]. Recently, solitons have also been revealed in more extended atomic structures on Si(100) [27]. Density functional theory (DFT) calculations seem to indicate that finite systems are not subjected to large electron-vibration effects because antiferromagnetic ordering counterbalances the atomic distortions [30].

These 1-D correlation effects have detrimental consequences in the wires' electron transport properties [31], reducing the possibilities of dangling-bond wires as alternative interconnects to free-standing [13, 32] or embedded [33] nanowires.

In order to remedy these problems, two strategies are reviewed here. Surface band engineering strategies [34] can be actively pursued by creating wires with given geometries that change the hybridization-to-correlation ratios leading to the closing of electronic gaps and hence to good conductance properties.

A second strategy is to use surface doping. When DB's are created, there is an attractive potential to dopants that approach the DB wires and modify their electronic structure leading to interesting transport properties [35]. Surprisingly, dopants greatly reduce the gap, which leads to quasi-metallic wires. A side effect is that as they stabilize the magnetic ordering in the wires, the wire transport becomes spin polarized. This leads to very good spin-filtering properties for the DB wires going beyond simple interconnects. Hence, DB wires will not only transport charge and spin, but will make sure that the spin orientation is preserved.

2 Computational Scheme

First-principles calculations are based on density functional theory (DFT) as implemented in Siesta [36, 37]. Calculations have been carried out with the GGA functional in the PBE form, [38] Troullier-Martins pseudopotentials, [39] and a double-ζ polarized basis set of finite-range numerical pseudoatomic orbitals for the valence wave functions [40]. A slab geometry with eight silicon layers and a 2×2 surface unit cell was used with a $5 \times 3 \times 1$ k-point sampling of the Brillouin zone. The geometrical relaxation has been performed until forces on the upper four layers were smaller than 0.04 eV/Å. The transport of infinite periodic wires is given by the number of contributing electronic bands at a given electron energy. For a qualitative evaluation of the conducting quality of the considered wires, the Landauer-Büttiker expression for electron currents is used [41].

3 Band Engineering in One Dimension

Let us consider an infinitely long antiferromagnetically coupled wire (AFM wire) formed by a single-row of dangling bonds on a H-passivated Si(100) surface [19, 21, 22, 24]. As for the electron transmission of previously studied finite wires [31], the strong electronic correlations leading to the AFM wire open an important gap that prevents good low-bias electron transport. It is interesting to notice that the AFM solution cannot distinguish between spins, and it is thus unable to perform any spin selective transport. This is not the case in the wires we will review in the second part of this chapter.

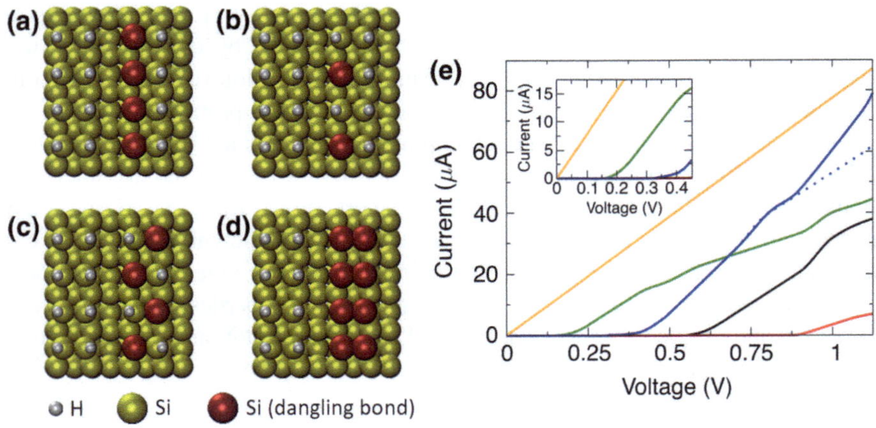

Fig. 1 Atomic structure of **a** the single-row antiferromagnetic wire (AFM), **b** the half-row, **c** the zigzag and **d** the double-row dangling-bond (dangling-bond) wires. H atoms are depicted in *white*, Si atoms in *yellow* and *red* when holding a dangling-bond. **e** Corresponding *I–V* curves: single-row AFM dangling-bond wire (*black*), half-row (*red*), zigzag (*green*) and double-row (*blue*). In this last case, a sizable bulk contribution to the current starts in the presented bias window. In *dashed lines*, the surface-state contribution of the double-row wire is shown. In full line, the full current is presented. The *yellow line* is the *I–V* of a one-channel metallic wire, such as a free-standing monoatomic gold wire

In order to reduce the electronic correlations, we decouple the dangling bond along the wire direction. Hence, we consider a new wire consisting of an alternating dangling-bond structure, Fig. 1b. Despite the reduced interaction along the wire, correlations are still important and the wire presents a ferromagnetically (FM) coupled ground state. The second strategy is to moderately increase the lateral size of the wires in order to have dangling bonds on both sites along the dimer row. This brings us two new wires: (i) a zigzag wire where the dangling bond changes site with adjacent dimers and (ii) a double-row wire where both dimer sites are dangling bonds. Figure 1 shows the four wires considered in this work.

The half-row wire contains a periodic potential that will modulate the surface band. This periodic potential is due to the alternated passivation of dangling bonds along the wire. This type of structures have been used to modify the Au(111) surface state by one-dimensional arrays of adsorbed molecules [42]. However, there are no induced states in the AFM gap, and a ferromagnetic (FM) ground state is found with the consequence of enhancing the electronic gap. Now, the FM character of the system is revealed in the electronic transmission function that shows two plateaux that are spin polarized.

We can reduce correlation effects causing the AFM and FM solutions by increasing the lateral dimension of the engineered surface states. This can be achieved by displacing one of the passivating H atoms of the AFM wire in order to create a zigzag dangling-bond structure that reduces correlations by decreasing confinement. The ground state is not magnetic meaning that indeed the electronic correlations are reduced. However, an important buckling of the surface dimers is found, exceeding the clean surface dimer buckling by ~0.1 Å in the vertical dimer displacement. This dimer distortion leads to a Peierls-like opening of a gap in the dangling-bond surface state. Luckily, the electronic gap is considerably smaller. It has gone down to ~0.2 eV from ~0.7 eV for the AFM wire and from ~0.9 eV for the half-row one.

One objection that can be made is that these structures will be difficult to obtain by atom manipulation because the passivating atoms will move back to their original DB's. Our evaluation of the H-diffusion barriers is about 1 eV for different diffusion paths along dangling bonds [24]. Hence, even though the zigzag wire is 110 meV/dimer higher in energy than the single-row wire, the proposed structure should be achievable with present scanning-probe manipulation means.

Evaluating the actual transport properties of these zigzag wires, we find that the conduction is comparable to the one of a good conductor such as free-standing monoatomic gold wires [43]. Figure 1e shows the *I–V* of a monoatomic gold wire which yields comparable currents as the one carried by the zigzag wire (within a factor of two).

Finally, a wire formed by a row of dangling-bond dimers, Fig. 1d, should present the lowest electronic correlation. And indeed, the double-row wire spin-polarized solution is 290 meV/dimer higher in energy than the non-magnetic solution. The distorted system adopts a buckled geometry similar to the case of the non-passivated Si(100) surface, Fig. 1a. In the buckled configuration, the two Si atoms of one dimer move 0.79 Å in the vertical direction. In addition, the subsurface atoms undergo a slight dimerization of 0.24 Å along the wire's direction. As for the Si(100) surface case, the π-backbond is perturbed by the distortion and a gap opens in the electronic structure. The consequence in the transmission is again two surface-state transmission bands, one for occupied and the other one for empty states, similar to the zigzag case.

Despite these good properties, the large perturbation of the DB in the electronic structure of the full systems leads to considerably mixing with bulk states. Hence, from 0.8 V on, the current is composed of surface-state and bulk-state contributions which is very bad for a surface interconnect. Figure 1e (blue line) shows the estimated surface contribution in dashed lines and the total current in full line. Beyond 0.8 V, there is an increasing bulk contribution to the current.

This work shows that crafting wires with a marked zigzag structure is the winning strategy for creating 1-D-like surface interconnects that otherwise would be highly correlated systems and hence poor electron conductors. A brute-force increase of the DB in a surface interconnect leads to considerably mixing with bulk states with the effect of injecting electrons in bulk states instead of carrying them along the surface. Hence, 1-D-like structures are probably the way to go in surface interconnect technology but with special geometries that favor the electronic transmission along the surface.

4 Band Engineering Through Charge Donors and Acceptors

A parallel strategy to produce wires of given geometries, including surface impurities with marked donor/acceptor character, is very appealing. Mainly, there is more technical flexibility using atomic manipulation techniques due to the lower number of manipulations needed to introduce these dopants close to the wire. Indeed, our calculations show that dopants are attracted to the reactive DB sites and there is a net diffusion of impurities from the bulk material to the area of the DB wire [35].

The impact of dopants on the current carried by DB wires depends critically on their distribution with respect to the surface. Hence, our first goal was to study the surface segregation of these defects. Broadly speaking, impurities prefer to be closer to the surface because the introduced strain can be released more easily. This is a well-known behavior in unpassivated Si surfaces [44, 45] and in Si nanowires [14, 46, 47]. Our calculations show the same trend in the H-passivated Si(100) surface. For both B and P, nearly 150 meV are gained in the most stable surface position (see Fig. 2b and Table 1). As the dopant gets closer to the surface, the formation energy becomes much more site-specific, with sites D, E and F being favored as a general rule. These differences vanish quickly moving away from the

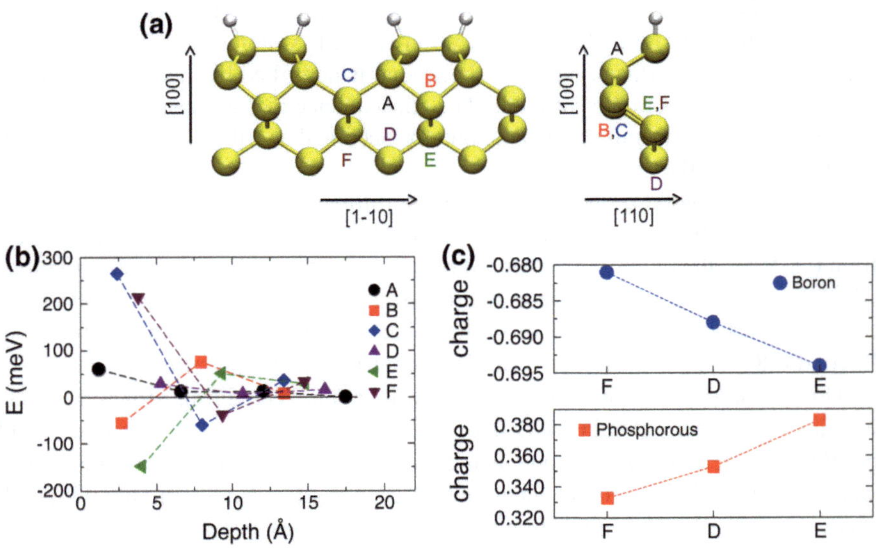

Fig. 2 **a** Atomic structure of the H-passivated Si(100) surface. The different lattice sites for B and P are indicated. **b** Formation energies with respect to the depth of the dopant (meV) for boron in Si close to the H-passivated Si(100) surface. At the bulk limit, all the sites A, B, C, D, E, and F are equivalent. The corresponding formation energy is taken as a reference. The formation energies become site-specific close to the surface. **c** Calculated Mulliken net charges of the boron and phosphorous impurities depending on the site substituted. *Dashed lines* are a guide to the eye

Table 1 Formation energies (in meV, see text) for sites D, E, and F with and without DB wires

	D	E	F
B@H-passivated	29	−148	214
B@NM	−14	−415	−12
B@AFM	−10	−110	−123
P@H-passivated	−61	−145	105
P@NM	−237	−488	−57
P@AFM	−329	−539	−180

The notations X@H-passivated, X@NM and X@AFM stands for a system with the dopant X (X = B, P) below a H-passivated surface, a NM and AFM wire, respectively

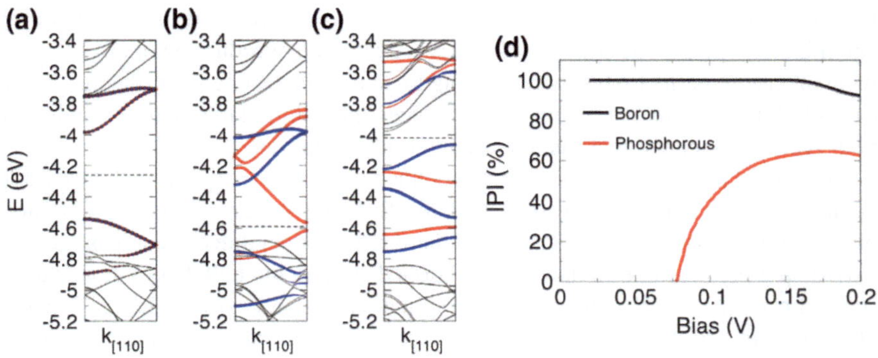

Fig. 3 **a–c** Electronic bands evaluated for a periodic magnetic nanowire with a pure Si substrate, B-doped Si substrate and P-doped Si substrate, respectively. The *red* and *blue circles* indicate the weight of the DB in the majority and minority spin bands, respectively. The *dashed lines* indicate the position of the Fermi energy. **d** Spin polarization of the current for the B-doped system (*black*) and the P-doped one (*red*). The curves are not defined for biases below the band gaps, since the wires hold no current. B-doped systems show 100% polarization for biases below 0.17 V. For larger biases, bulk contributions deteriorate the spin polarization. In the case of P-doped system, the bulk bands contribution starts as early as 0.09 eV, leading to a lower spin polarization. For biases lower than 0.17 V, the B-doped DB wire drawn on H-passivated Si(100) system is a perfect non-leaking spin-filtering surface interconnect

surface and are within the numerical accuracy of the calculation at 15 Å from the surface. In the following, we limit our discussion to the case of a substitution at sites D, E and F which appears to be more stable than those at A, B and C sites.

The presence of DB wires reduces the formation energy of substitutional impurities, B or P, as compared with bulk values, see Table 2. This is in agreement with previous results on Si NWs, [46, 48] where dopants are found to form electrically inactive complexes with isolated DBs. Also here, the tendency to surface segregation is more pronounced with P, as observed in the case of Si NWs and unpassivated Si(100) surface [46, 48].

Figure 3 shows the band structure of (a) an undoped AFM wire, (b) an AFM wire with a substitutional B atom in its most stable configuration (F site) and (c) an

Table 2 Energy differences (in meV/DB) between NM and AFM configurations for top sites D, E, and F

B@D	B@E	B@F	P@D	P@E	P@F
10	22	16	20	15	21

In the absence of dopant, this difference is −5 meV/DB [24]

AFM wire with a substitutional P atom in its most stable configuration (E site). Due to the injection of charge from the dopants, there is a considerably shifting of the Fermi energy in the direction dictated by the donor/acceptor character of the dopant. The undoped AFM wire, Fig. 3a, shows two surface states where the two spins overlap due to the AFM character of the electronic structure. The B-doped system, Fig. 3b, displays a splitting of bands according to spin (red and blue for majority and minority spins). The splitting of bands is due to the introduction of an extra spin in an otherwise perfect AFM wire, leading to an unbalanced number of spins. In the case of P doping, the extra spin also produces the spin polarization of bands. For both types of dopants, there is a substantial gap reduction.

We obtain that the spin polarization, defined as $P = (I_{\uparrow} - I_{\downarrow})/(I_{\uparrow} + I_{\downarrow})$, is 100% polarization for biases below 0.17 and for B-doped systems, Fig. 3d. For large biases, bulk bands start contributing and there is a considerable current leakage with loss of spin polarization. In the case of P-doped system, the bulk bands contribution starts already at 0.09 eV, and the overall spin polarization is lower. Hence, B-doped DB wires are perfect spin-filters thanks to the absence of current leakage and to the perfect spin polarization for low-bias applications.

5 Conclusions

Crafting DB wires on Si(100)-H is an appealing concept for the creation of surface interconnects that remains on the nanoscale. However, their large electron confinement leads to strong correlation effects such as Peierls distortions and magnetic ground states with the consequence of electronic band gaps. This has detrimental consequences in the possible electron transport and therefore in the capabilities of these wires functioning as surface interconnects. We explored these issues proposing new strategies to create wires that are good interconnects. These strategies are divided into two. The first one is the use of enhanced atomic manipulation tools that permit us to design wires of complex geometrical structures. The second one is the inclusion of dopants in the design of wires.

When the wires depart from a linear 1-D structure, the correlation effects are greatly reduced. We have found that zigzag like wires manage to reduce the correlations to levels where the electronic structure approaches metallic solutions and hence the electronic transmission becomes comparable with the one of a metallic wire of the same dimensions. Increasing the density of DB in one of these crafted wires is, however, not a good strategy because the stronger perturbation caused by the DB in the electronic structure of the full system leads to considerable mixing of surface and bulk electronic states. As a consequence, the electronic states are not confined to the surface any longer and bulk leakage from the wires takes place.

The second strategy consists in doping the systems where wires are crafted. This turns out to be energetically favorable because the dopants will place themselves close to the DB wires, changing their conduction properties. The main effect of the dopants is to close the band gaps of the wires, because of the stabilization of magnetic solutions induced by the dopants. As a consequence, transport becomes spin polarized through the wires, leading to the creation of spin-filtering interconnects.

Acknowledgements The authors would like to thank the European Union Integrated Project AtMol for financial support.

References

1. Joachim, C., Gimzewski, J.K., Aviram, A.: Electronics using hybrid-molecular and mono-molecular devices. Nature **408**, 541–548 (2000)
2. van Ruitenbeek, J.M., Alvarez, A., Pineyro, I., Grahmann, C., Joyez, M., Esteve, D., Urbina, C.: Adjustable nanofabricated atomic size contacts. Rev. Sci. Instrum. **67**, 108 (1996)
3. Park, H., Lim, A., Alivisatos, A., Park, J., Euen, P.: Fabrication of metallic electrodes with nanometer separation by electromigration. Appl. Phys. Lett. **75**, 301 (1999)
4. Eigler, D.M., Schweizer, E.K.: Positioning single atoms with a scanning tunnelling microscope. Nature **344**, 524 (1990)
5. Lopinski, G.P., Wayner, D.D.M., Wolkow, R.A.: Self-directed growth of molecular nanostructures on silicon. Nature **406**, 48 (2000)
6. Dappe, Y.J., González, C., Bulou, H., Berndt, R.: Charge injection through single and double carbon bonds. Nano Lett. **11**, 3142–3146 (2011)
7. Kane, B.E.: A silicon-based nuclear spin quantum computer. Nature **393**, 133 (1998)
8. Gimzewski, J.K., Joachim, C.: Nanoscale science of single molecules using local probes. Science **283**, 1783 (1999)
9. Hallam, T., Reusch, T.C.G., Oberbeck, L., Curson, N.J., Simmons, M.Y.: Scanning tunneling microscope based fabrication of nano- and atomic scale dopant devices in silicon: the crucial step of hydrogen removal. J. Appl. Phys. **101**, 034305 (2007)
10. Fuhrer, A., Fuechsle, M., Reusch, T.C.G., Weber, B., Simmons, M.Y.: Atomic-scale, all epitaxial in-plane gated donor quantum dot in silicon. Nano Lett. **9**, 707 (2009)
11. Fuechsle, M., Mahapatra, S., Zwanenburg, F.A., Friesen, M., Eriksson, M.A., Simmons, M.Y.: Spectroscopy of few-electron single-crystal silicon quantum dots. Nat. Nanotech. **5**, 502 (2010)
12. Fuechsle, M., Miwa, J.A., Mahapatra, S., Ryu, H., Lee, S., Warschkow, O., Hollenberg, L.C.L., Klimeck, G., Simmons, M.Y.: A single-atom transistor. Nat. Nanotech. **7**, 242 (2012)
13. Cui, Y., Lieber, C.M.: Functional nanoscale electronic devices assembled using silicon nanowire building blocks. Science **291**, 851–853 (2001)
14. Rurali, R.: Colloquium: structural, electronic, and transport properties of silicon nanowires. Rev. Mod. Phys. **82**, 427–449 (2010)
15. Watanabe, S., Ono, Y.A., Hashizume, T., Wada, Y.: Theoretical study of atomic and electronic structures of atomic wires on an H-terminated Si(100)2×1 surfaces. Phys. Rev. B **54**, R17308 (1996)

16. Hitosugi, T., Heike, S., Onogi, T., Hashizume, T., Watanabe, S., Li, Z.-Q., Ohno, K., Kawazoe, Y., Hasegawa, T., Kitazawa, K.: Jahn-Teller distortion in dangling-bond linear chains fabricated on a hydrogen-terminated Si(100)-2×1 surface. Phys. Rev. Lett. **82**, 4034 (1999)
17. Doumergue, P., Pizzagalli, L., Joachim, C., Altibelli, A., Baratoff, A.: Conductance of a finite missing hydrogen atomic line on Si(001)-2×1-H. Phys. Rev. B **59**, 15910–15916 (1999)
18. Cho, J.-H., Kleinman, L.: Nature of lattice distortion in one-dimensional dangling-bond wires on Si and C. Phys. Rev. B **66**, 235405 (2002)
19. Bird, C.F., Bowler, D.R.: A spin-polarised first principles study of short dangling bond wires on Si(001). Surf. Sci. **531**, L351–L355 (2003)
20. Soukiassian, L., Mayne, A.J., Carbone, M., Dujardin, G.: Atomic wire fabrication by STM induced hydrogen desorption. Surf. Sci. **528**, 121 (2003)
21. Lee, J.Y., Cho, J.-H., Zhang, Z.: Quantum size effects in competing charge and spin orderings of dangling bond wires on Si(001). Phys. Rev. B **80**, 155329 (2009)
22. Lee, J.-H., Cho, J.-H.: Instability of one-dimensional dangling-bond wires on H-passivated C (001), Si(001), and Ge(001) surfaces. Surf. Sci. **605**, L13–L15 (2011)
23. Bianco, F., Owen, J.H.G., Köster, S.A., Mazur, D., Renner, C., Bowler, D.R.: Endotaxial Si nanolines in Si(001):H. Phys. Rev. B **84**, 035328 (2011)
24. Robles, R., Kepenekian, M., Monturet, S., Joachim, C., Lorente, N.: Energetics and stability of dangling-bond silicon wires on H passivated Si(100). J. Phys. Condens. Matter **24**, 445004 (2012)
25. Hitosugi, T., Heike, S., Onogi, T., Hashizume, T., Watanabe, S., Li, Z., Ohno, K., Kawazoe, Y., Hasegawa, T., Kitazawa, K.: Jahn-Teller distortion in dangling-bond linear chains fabricated on a hydrogen-terminated Si(100)-2×1 Surface. Phys. Rev. Lett. **82**, 4034 (1999)
26. Soukiassian, L., Mayne, A.J., Carbone, M., Dujardin, G.: Atomic wire fabrication by STM induced hydrogen desorption. Surf. Sci. **528**, 121–126 (2003)
27. Lai, M.Y., Chou, J.P., Utas, O.A., Denisov, N.V., Kotlyar, V.G., Gruznev, D., Matetsky, A., Zotov, A.V., Saranin, A.A., Wei, C.M., Wang, Y.L.: Broken even-odd symmetry in self-selection of distances between nanoclusters due to the presence or absence of topological solitons. Phys. Rev. Lett. **106**, 166101 (2011)
28. Bowler, D.R., Fisher, A.J.: Small polaron formation in dangling-bond wires on the Si(001) surface. Phys. Rev. B **63**, 035310 (2000)
29. Bird, C.F., Fisher, A.J., Bowler, D.R.: Soliton effects in dangling-bond wires on Si(001). Phys. Rev. B **68**, 115318 (2003)
30. Lee, J.Y., Cho, J., Zhang, Z.: Quantum size effects in competing charge and spin orderings of dangling bond wires on Si(001). Phys. Rev. B **80**, 155329 (2009)
31. Kepenekian, M., Novaes, F.D., Robles, R., Monturet, S., Kawai, H., Joachim, C., Lorente, N.: Electron transport through dangling-bond silicon wires on H passivated Si(100). J. Phys. Condens. Matter **25**, 025503 (2013)
32. Rurali, R.: Colloquium: structural, electronic, and transport properties of silicon nanowires. Rev. Mod. Phys. **82**, 427–449 (2010)
33. Weber, B., Mahapatra, S., Ryu, H., Lee, S., Fuhrer, A., Reusch, T.C.G., Thompson, D.L., Tee, W.C.T., Klimeck, G., Hollenberg, L.C.L., Simmons, M.Y.: Ohm's law survives to the atomic scale. Science **335**, 64 (2012)
34. Kepenekian, M., Robles, R., Joachim, C., Lorente, N.: Surface-state engineering for interconnects on H-passivated Si(100). Nano Lett. **13**, 1192–1195 (2013)
35. Kepenekian, M., Robles, R., Rurali, R., Lorente, N.: Spin transport in dangling-bond wires on doped H-passivated Si(100). Nanotechnology **25**, 465703 (2014)
36. Soler, J.M., Artacho, E., Gale, J.D., García, A., Junquera, J., Ordejón, P., Sánchez-Portal, D.: The SIESTA method for ab initio order-N materials simulation. J. Phys. Condens. Matter **14**, 2745–2779 (2002)
37. Artacho, E., Anglada, E., Diéguez, O., Gale, J.D., García, A., Junquera, J., Martin, R.M., Ordejón, P., Pruneda, J.M., Sánchez-Portal, D., Soler, J.M.: The SIESTA method; developments and applicability. J. Phys. Condens. Matter **20**, 064208 (2008)

38. Perdew, J.P., Burke, K., Ernzerhof, M.: Generalized gradient approximation made simple. Phys. Rev. Lett. **77**, 3865 (1996)
39. Troullier, N., Martins, J.L.: EfFicient pseudopotentials for plane-wave calculations. Phys. Rev. B. **43**, 1993 (1991)
40. Artacho, E., Sánchez-Portal, D., Ordejón, P., García, A., Soler, J.M.: Linear-scaling ab-initio calculations for large and complex systems. Phys. Stat. Sol. B. **215**, 809 (1999)
41. Datta, S.: Electronic Transport in Mesoscopic Systems. Cambridge University Press (2007)
42. Gonzalez-Lakunza, N., Fernández-Torrente, I., Franke, K.J., Lorente, N., Arnau, A., Pascual, J.I.: Formation of dispersive hybrid bands at an organic-metal interface. Phys. Rev. Lett. **100**, 156805 (2008)
43. Pascual, J.I., Méndez, J., Gómez-Herrero, J., Baró, A.M., García, N., Binh, V.T.: Quantum contact in gold nanostructures by scanning tunneling microscopy. Phys. Rev. Lett. **71**, 1852–1855 (1993)
44. Luo, X., Zhang, S.B., Wei, S.-H.: Understanding ultrahigh doping: the case of boron in silicon. Phys. Rev. Lett. **90**, 026103 (2003)
45. Centoni, S.A., Sadigh, B., Gilmer, G.H., Díaz de la Rubia, T., Musgrave, C.B.: First-principles calculation of free Si(100) surface impurity enrichment. Appl. Phys. Lett. **87**, 232101 (2005)
46. Fernández-Serra, M.V., Adessi, C., Blase, X.: Surface segregation and backscattering in doped silicon nanowires. Phys. Rev. Lett. **96**, 166805 (2006)
47. Peelaers, H., Partoens, B., Peeters, F.M.: Formation and segregation energies of B and P doped and BP codoped silicon nanowires. Nano Lett. **6**, 2781 (2006)
48. Fernández-Serra, M.V., Adessi, C., Blase, X.: Conductance, surface traps, and passivation in doped silicon nanowires. Nano Lett. **6**, 2674–2678 (2006)

Band Engineering of the Si(001):H Surface by Doping with P and B Atoms

Hiroyo Kawai, Kuan Eng Johnson Goh, Mark Saeys and Christian Joachim

Abstract Si(001):H surfaces doped with phosphorus (P) and with boron (B) were explored as conducting contact pads for a dangling bond (DB) wire. Heavily doped Si(001):H surface patches are proposed as an alternative to metallic nano-islands or as an intermediate scale between the DB wire and the metallic islands to precisely contact a DB wire. Our calculations show that patches of B dopants incorporated in Si(001):H surface introduce states in the surface band gap with a significant dispersion along the dimer rows and which overlap well with the DB wire band states, demonstrating the possibility of using patches a B-doped Si(001):H as viable and robust contact pads for DB wires.

1 Introduction

The fabrication of atomic scale structures has been widely explored due to potential applications in atomic scale electronic devices [1]. One possible way to fabricate such devices on a surface is by adsorbing a molecule on a passivated semiconductor surface and interconnecting the molecule with surface atomic wires created by atom-by-atom removal of the passivating atoms from the semiconductor surface [2].

H. Kawai (✉) · K.E.J. Goh
Institute of Materials Research and Engineering,
2 Fusionopolis Way, Innovis, #08-03, Singapore 138634, Singapore
e-mail: kawaih@imre.a-star.edu.sg

M. Saeys
Laboratory for Chemical Technology, Ghent University,
Technologiepark 914, 9052 Ghent, Belgium

C. Joachim
GNS & MANA Satellite, CEMES-CNRS,
29 rue J. Marvig, 31055 Toulouse Cedex, France

C. Joachim
International Center for Materials Nanoarchitectronics (WPI-MANA), National Institute for Materials Science (NIMS), 1-1 Namiki, Tsukuba, Ibaraki 305-0044, Japan

M. Kolmer and C. Joachim (eds.), *On-Surface Atomic Wires and Logic Gates*,
Advances in Atom and Single Molecule Machines,
DOI 10.1007/978-3-319-51847-3_6

The depassivated sites introduce dangling bond (DB) states within the surface band gap, and a line of interacting DBs can be used as a conducting or semiconducting wire [2–4]. On the single-molecule scale, molecules have been designed to perform different types of logic functions. For example, a single-molecule NOR logic function has been demonstrated experimentally using the trinaphthylene molecule, where the quantum states were manipulated by contacting the naphthyl branches of the molecule with Au atoms [5]. In addition, electron transport calculations for a structurally similar molecule, di-(9-amino,10-hydro-anthracene)[a,c]naphthacene (DAN), interconnected with DB wires on a Si(001):H surface, demonstrated that a molecule interconnected with DB wires can function as an OR logic gate [2]. Another way to fabricate atomic scale devices is by directly exploiting the gap states introduced by the DBs, without using a molecule. In theory, it is possible to design atomic scale Boolean logic gates with two inputs and one output on a Si (001):H surface using only these DB states [2, 6].

For any atomic scale device on the surface, one of the remaining challenges lies in the interconnection of the devices. It was proposed that the atomic devices can be interconnected by precisely contacting them with atomic scale DB wires, which are then in turn contacted by larger metallic nano-pads, such as Au islands physisorbed onto the passivated surface [2, 6, 7]. An example of a DB wire contacted to Au nano-pads is illustrated in Fig. 1a.

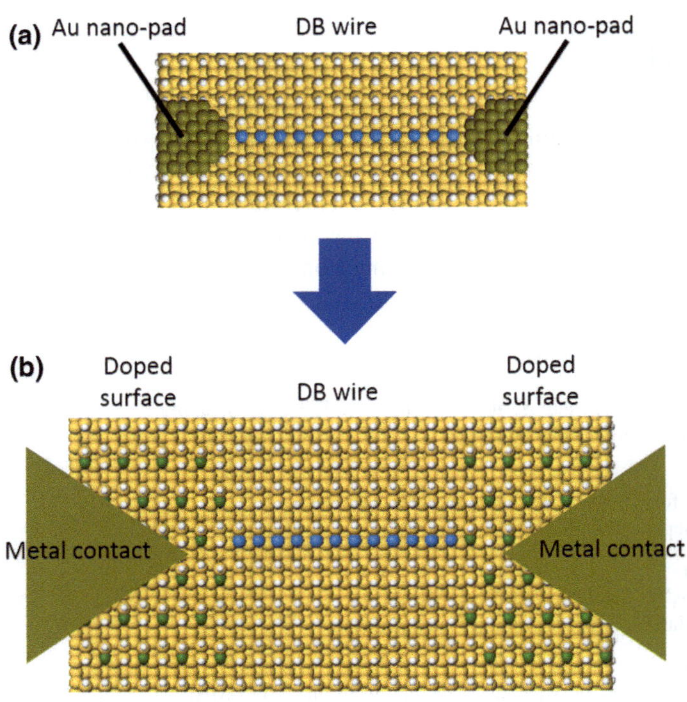

Fig. 1 Atomic structures of a DB wire contacted by **a** Au nano-pads and **b** by doped surface patches and subsequently by metal contacts

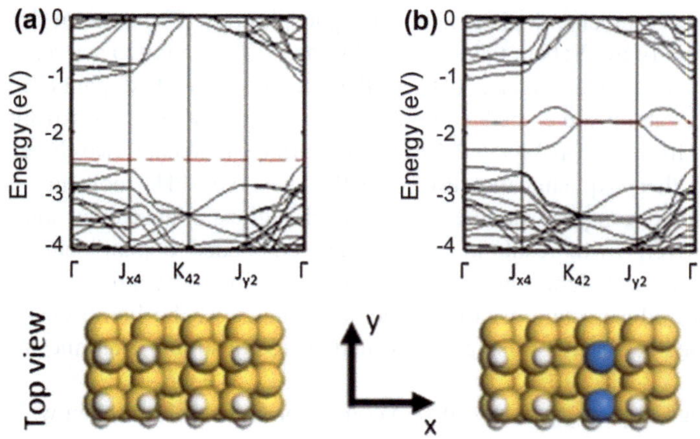

Fig. 2 Band structures and atomic structures of **a** the fully hydrogenated Si(001):H surface and **b** the Si(001):H surface with an ideal DB wire crated along a dimer row. Notice the 0.8 eV dispersion of the DB wire

An ideal non-relaxed line of DBs along the dimer reconstruction rows creates a DB-derived band with a dispersion of about 0.8 eV due to the strong coupling of neighboring DB states. Such a line can hence function as a pseudo-ballistic wire [3, 4] (Fig. 2). A good contact between the DB wire and the metallic nano-pads is important for maximum conductance. As the DB wire is essentially single atom wide, the precise positioning of the larger Au-pads is crucial. However, since these metallic nano-pads are only physisorbed on the surface, they are not well anchored and may shift. Even small changes in the orientation of the nano-pad relative to the DB wire significantly affect the conductance of the atomic scale contact [8]. In addition, it is challenging to position the metallic nano-pads with atomic precision to contact a DB wire of only a few nanometers. Less than perfect positioning of the nano-pads may result in large leakage currents between nano-pads and hence seriously hamper the function of the atomic devices.

As we illustrate in this chapter, one way to address this interconnection challenge is to heavily dope patches of the Si(001):H surface. Doping of the surface potentially introduces channels within the surface band gap. The doped surface patches can then act as a robust intermediate scale to precisely contact a DB wire. Metallic contacts can then still contact the DB wire indirectly via the doped patches as illustrated in Fig. 1b, reducing the challenges of precisely positioning the metallic nano-pads. Indeed, unlike the physisorbed metallic nano-pads, the doped surface patches remain fixed and precisely defined, unless the surface is modified. In addition, since the metallic pads now contact the doped surface patches instead of the DB wire, atomically precise positioning of the metallic pads is no longer required, and the distance between the metallic pads can increased, reducing the effect of leakage currents.

Atomically precise doping of Si with phosphorous (P) has been demonstrated experimentally by Weber et al. using STM-based lithography to activate the surface to be doped, and gas phase PH_3 as the source of the P dopants [9, 10]. The phosphorus atoms are embedded within the Si crystal by first depassivating rows of Si atoms from a H-passivated Si(001):H surface to create two-dimer-row-wide lines, and then exposing the surface to PH_3 gas. The PH_3 molecules selectively adsorb and decompose on the depassivated regions. Subsequent annealing and epitaxial overgrowth result in a 2D planar embedded dopant density of −0.25 monolayers [9]. A tight-binding theoretical study of this structure showed that the P dopants introduce a number of conducting channels at the Fermi energy, demonstrating the possibility of using such structures to constructing atomic scale devices [9, 10].

Although doping processes of semiconductor surfaces have been widely studied [11–14], the electronic structure of surface dopants and the possible applications of surface dopants have rarely been discussed. In this chapter, different elements and different configurations are explored as possible conducting surface patches on the Si(001):H surface to interconnect atomic scale electronic devices. In order to use the doped surface patches to contact a DB wire, it is necessary that the dopants introduce states within the surface band gap at the energy range of the DB wire band, that those dopant states are well coupled with each other, in particular along the DB wire direction, and that the dopant states couple to the DB states. In the subsequent sections, the electronic properties of the Si(001):H surface doped with two different elements and for different configurations are analyzed and discussed. Our simulations suggest that patches of boron dopants could match all the requirements outlined above.

2 Phosphorus Dopants on a Si(001):H Surface

First, we explored the possibility of introducing phosphorus atoms as dopants for the Si(001):H surface. This was motivated by studies revealing that doping Si bulk with P atoms introduces conducting channels near the Fermi energy [9, 10]. Unlike these earlier studies, we introduced the P dopants as substituents of ***surface*** Si atoms of the Si(001):H surface. Similar to the earlier work, we used a doping density of 0.25 monolayers P due to limits introduced by the surface doping process based on dosing PH_3. To satisfy the valency of the surface P dopants, we also explored structures where the P surface atoms are bound with 1 or 2 H atoms. The surfaces were modeled as 5-layer (4 × 2) Si(100) slabs, and the slab surfaces were passivated with H atoms. Optimized surface geometries and band structures were calculated using density functional theory (DFT) with the Perdew–Burke–Ernzerhof (PBE) functional and a (3 × 7 × 1) Monkhorst Pack k-point grid [15] as implemented in the Vienna Ab initio simulation package (VASP) [16–19].

The surface band structures and atomic structures for P-doped Si(001):H surfaces are shown in Fig. 3a–c. The band structure of the Si(001):H surface doped

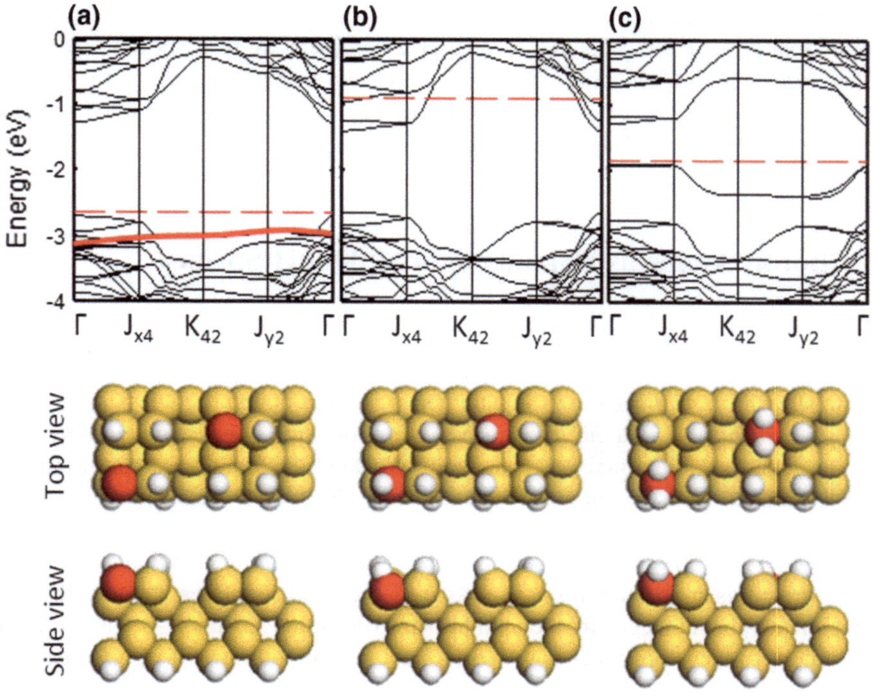

Fig. 3 Band structures and atomic structures of a P-doped Si(100):H. **a** Unsaturated P atoms, **b** P atoms saturated with 1 H atom (most stable structure), and **c** structure with 2 H atoms attached to each P dopant

with P atoms without H passivation is similar to the band structure of the fully hydrogenated Si(001):H surface, except for an additional P dopant band introduced at the valence band edge (VBE) (indicated in red in Fig. 3a). Since this P dopant state lies below the VBE and does not extend into the Si(100) surface band gap, this configuration cannot be used to contact the DB wire where the DB wire band lies within the band gap (Fig. 2b). When the surface P dopants are saturated with H atoms, the P dopant band at the VBE shifts all the way to the conduction band and shows a large dispersion.

As a result, the Fermi energy shifts into the conduction band, and some bands cross the new Fermi energy. However, since the Si(001):H surface band gap remains for this configuration and no states are introduced in the gap (Fig. 3b), this configuration does not satisfy the requirements to contact the DB wire. Saturating the surface P dopants with 2 H atoms introduces bands inside the surface band gap (Fig. 3c). However, these bands at VBE and CBE are not introduced by the P dopants, but rather by the Si atoms below the P atom, which are now not fully coordinated. The band dispersion along y-axis (along the dimer row) suggests that the new bands might introduce conducting channels along the dimer rows. However, the proposed structure does not satisfy the chemical valency of P and is hence highly unstable. Indeed, attaching a second H atom to the P dopants is unfavorable by 1 eV, and the second H is unlikely to bind to the P dopant. Therefore, while doping bulk Si by P atoms introduces conducting channels at Fermi energy, doping of the Si(001):H surface with P atoms will not introduce conducting patches to contact a surface DB wire.

3 Boron Atoms as Dopants on a Si(001):H Surface

Since the valency of P did not allow to find stable P-doped surface structures which introduce states within the Si(001):H surface band gap, we decided to explore boron as an alternative. Surface boron is saturated when it is bound to three Si atoms and the attachment of a H atom to a surface boron dopant is hence unfavorable. The dimer with a substitutional boron dopant is buckled down, and the boron atom is sp^2-hybridized with an empty *p*-derived state (Fig. 4). Different surface boron doping configurations were explored, keeping the dopant surface coverage at 0.25 monolayers (Fig. 4a–d). The structures, their relative stabilities and corresponding band structures are shown in Fig. 4. Based on the band structures, the transmission coefficients (being equal to the number of conducting channels because of the periodicity) along the dimer row were calculated for infinitely long B lines and compared with the number and the position of the channels introduced by a DB wire (Fig. 5).

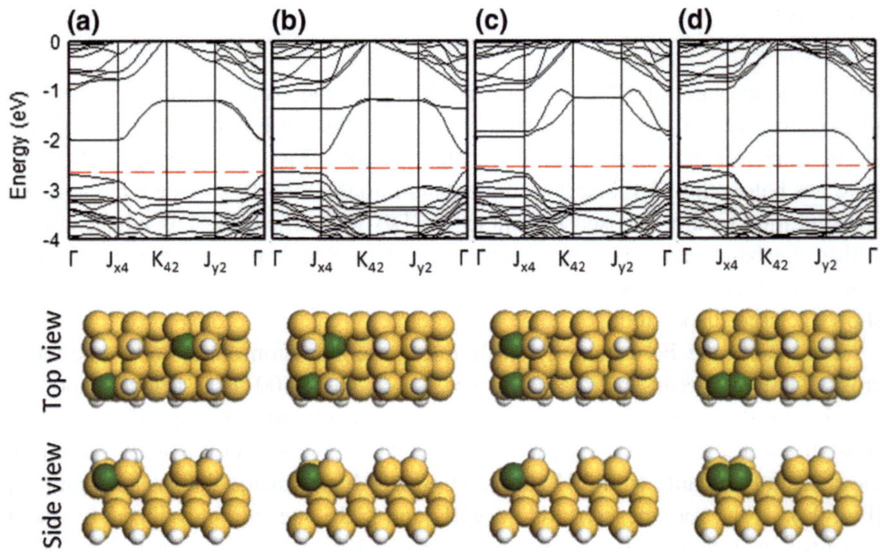

Fig. 4 **a–d** Band structure, atomic structure, and relative stability for four B surface doping configurations

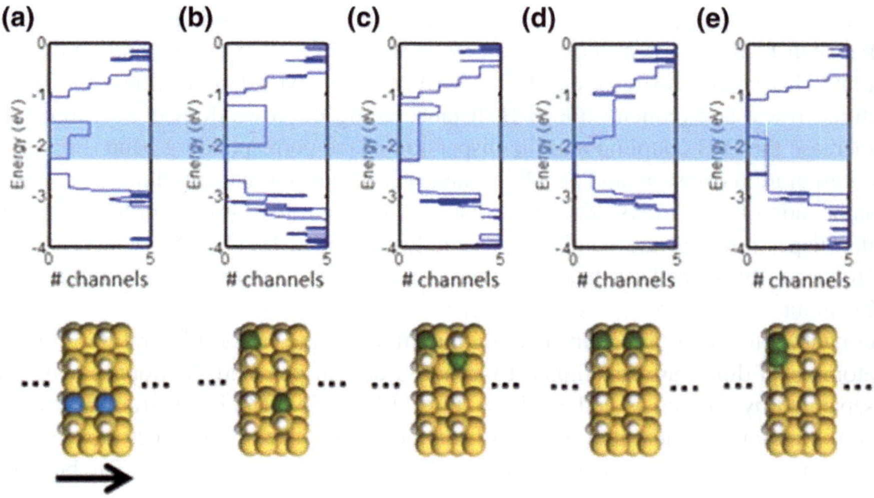

Fig. 5 Number of channels for an infinitely long **a** DB wire and **b–e** B dopants lines with different configurations along the surface dimer rows. The *black arrow* shows the direction of the transport. The energy range of DB wire band is highlighted to indicate possible overlap with the surface B-derived conduction channels

The band structure for a B-doped surface with a dopant in every other dimer shows that surface B dopants introduce bands inside the Si(001):H surface band gap, (Fig. 4a) and hence B-doped Si(001):H surface patches have the potential to function as contacts for a DB wire. The B-derived bands have a large dispersion of about 0.8 eV along the dimer row, similar to the dispersion of the DB wire band, even though the distance between B dopants is double the distance between Si DB in the DB wire. Each B atom in the unit cell introduces one conducting channel, and B atoms in different dimer rows interact only weakly. The large dispersion of the B-derived bands along the dimer rows indicates that B-derived states interact strongly along the dimer rows, even though the B dopants are separated by a H-passivated Si dimer. The large extension of the B-derived state along the dimer rows (but not between dimer rows) is consistent with our study of isolated Si DB on p-type Si(001):H [10]. Also on p-type Si, the Si DB atom is sp^2-hybridized and STM images and calculations demonstrate that the state derived from the empty Si DB p-orbital extends several Si dimers along the dimer rows [10] (see DB wire chapter for details).

When the B dopants are introduced in every dimer and on opposite sides along a dimer row, the interaction between the B dopants splits the bands (Fig. 4b). Although this interaction slightly increases the band dispersion, the splitting of the band creates an increased energy range with one channel and a small energy range with two channels. This may be less favorable than conducting channels for the first configuration (Fig. 4a). The structure where a B dopant is introduced in every dimer is also less stable than the structure where the dopants are more separated. The increase in the dispersion is much less than expected considering that the distance between B dopants is reduced by almost a factor of two in this second configuration. This again illustrates that the B-derived states extend significantly along the dimer rows, and reducing the B-B distance along the rows does not significantly increase the B-B coupling and the dispersion of the corresponding band. In a third configuration (Figs. 4c and 5d), B dopants are introduced in every dimer and on the same side of the dimers. The short distance between the dopants further increases the dispersion, and the band now resembles the band for the DB wire (Fig. 2b). However, this is the least stable configuration (less stable than the first configuration by about 140 meV per B atom). It may therefore be challenging to introduce boron dopants into every dimer in this configuration. In the last configuration, both Si atoms of a dimer are substituted by a boron dopant pair and the dopant pairs are separated by a H-passivated Si dimer (Figs. 4d and 5e). The band structure for this configuration appears quite different from the others: The boron dimer introduces a bonding band near the VBE and an anti-bonding band above the gap. The bonding boron band near the VBE falls above the Fermi energy and is empty, as expected from the B valency. The bonding band near the VBE edge provides 1 channel per B dimer and resembles the band for the the first structure (Fig.4a). This configuration is less stable than the first configuration by about 55 meV per B. The first configuration with the largest separation between boron atoms is hence both the most stable configuration and provides the most interesting band structure.

As shown in Fig. 5, all B dopant configurations considered here provide conducting channels within the energy range of DB wire band, showing the possibility of using B dopant patches on Si(001):H to contact a DB wire. To evaluate the contact conductance between the B patch and the DB wire, it is necessary to calculate transmission coefficients through B patch-DB wire-B-patch junctions (see Fig. 1b for example), which is beyond the scope of this work. In addition, it is necessary to study the actual doping process and the effect of metal contact pads on boron structural and electronic properties to evaluate the feasibility of fabricating the proposed structures.

4 Conclusions

We have evaluated the possibility of using P or B surface dopants to engineer surface-doped regions on a Si(001):H surface which can function as robust contact pads for DB wires. Our calculations show that for P dopants, only a structure where 2 H atoms are attached to the surface P dopant introduces states in the Si(001):H surface band gap which might overlap with the energy range of DB wire band. However, in this configuration, the P dopant is over-coordinated and hence unstable. The calculations hence suggest that unlike for bulk Si doping, surface P-doping is not a promising strategy to fabricate contact pads for DB wires. Inspired by this finding, we explored substitutional surface B doping. B dopants are saturated when they are bound to three Si atoms. Interestingly, B dopant-derived states fall inside the Si surface band gap and couple strongly along the dimer rows. Even for a dopant coverage of only 0.25 ML, corresponding to introducing a B dopant in every other dimer, and similar to dopant coverages that have been realized experimentally for P, B-derived states couple strong enough to create a 0.8 eV-wide conductive channel which overlaps well with the conducting channel of the DB wire. Other configurations where a B dopant was introduced in every dimer along a row were less stable and did not significantly increase the width of the B-derived conducting channel. Although further transport simulations are required to evaluate the viability of using B-doped patches to effectively contact DB wires, as well as the feasibility of fabricating the proposed patches, this work demonstrates the possibility of using boron dopants to alleviate challenges faced in the interconnection of atomic devices on surfaces.

Acknowledgements We acknowledge the Agency of Science, Technology, and Research (A*STAR) for funding provided through the Visiting Investigatorship Programme Atom Technology project 1021100972, and through the AtMol integrated project contract number 270028 from the European Commission. We also acknowledge the A*STAR Computational Resource Centre (A*CRC) for computational resources and support.

References

1. Joachim, C., Gimzewski, J.K., Aviram, A.: Electronics using hybrid-molecular and mono-molecular devices. Nature **408**, 541 (2000). doi:10.1038/35046000
2. Ample, F., Duchemin, I., Hliwa, M., Joachim, C.: Single OR molecule and OR atomic circuit logic gates interconnected on a Si(100)H surface. J. Phys.: Condens. Matter **23**, 125303 (2011). doi:10.1088/0953-8984/23/12/125303
3. Kawai, H., Yeo, Y.K., Saeys, M., Joachim, C.: Conductance decay of a surface hydrogen tunneling junction fabricated along a Si(001)-(2×1)-H atomic wire. Phys. Rev. B **81**, 195316 (2010). doi:10.1103/PhysRevB.81.195316
4. Robles, R., Kepenekian, M., Monturet, S., Joachim, C., Lorente, N.: Energetics and stability of dangling-bond silicon wires on H passivated Si(100). J. Phys.: Condens. Matter **24**, 445004 (2012). doi:10.1088/0953-8984/24/44/445004
5. Soe, W.-H., Manzano, C., Renaud, N., de Mendoza, P., De Sarkar, A., Ample, F., Hliwa, M., Echavarren, A.M., Chandrasekhar, N., Joachim, C.: Manipulating molecular quantum states with classical metal atom inputs: demonstration of a single molecule NOR logic gate. ACS Nano **5**, 1436 (2011). doi:10.1021/nn1033899
6. Kawai, H., Ample, F., Wang, Q., Yeo, Y.K., Saeys, M., Joachim, C.: Dangling-Bond logic gates on a Si(100)-(2×1)-H surface. J. Phys.: Condens. Matter **24**, 095011 (2012). doi:10.1088/0953-8984/24/9/095011
7. Joachim, C., Martrou, D., Rezeq, M., Troadec, C., Deng, J., Chandrasekhar, N., Gauthier, S.: Multiple atomic scale solid surface interconnects for atom circuits and molecule logic gates. J. Phys.: Condens. Matter **22**, 084025 (2010). doi:10.1088/0953-8984/22/8/084025
8. Otalvaro, D.M., Duchemin, I., Saeys, M., Joachim, C.: Calculation of the conductance of a finite atomic line of sulfur vacancies created on a molybdenum disulfide surface. Phys. Rev. B **77**, 205429 (2008). doi:10.1103/PhysRevB.77.205429
9. Weber, B., Mahapatra, S., Ryu, H., Lee, S., Fuhrer, A., Reusch, T.C.G., Thompson, D.L., Lee, W.C.T., Klimeck, G., Hollenberg, L.C.L., Simmons, M.Y.: Ohm's law survives to the atomic scale. Science **335**, 64 (2012). doi:10.1126/science.1214319

10. Ryu, H., Lee, S., Weber, B., Mahapatra, S., Hollenberg, L.C.L., Simmons, M.Y., Klimeck, G.: Atomistic modeling of metallic nanowires in silicon. Nanoscale **5**, 8666 (2013). doi:10.1039/c3nr01796f
11. Goh, K.E.J., Oberbeck, L., Butcher, M.J., Curson, N.J., Rueß, F.J., Simmons, M.Y.: Comparison of GaP and PH_3 as dopant sources for STM-based device fabrication. Nanotechnology **18**, 065301 (2007). doi:10.1088/0957-4484/18/6/065301
12. Wilson, H.F., Warschkow, O., Marks, N.A., Curson, N.J., Schofield, S.R., Reusch, T.C.G., Radney, M.W., Smith, P.V., McKenzie, D.R., Simmons, M.Y.: Thermal dissociation and desorption of PH_3 on Si(001): a reinterpretation of spectroscopic data. Phys. Rev. B **74**, 195310 (2006). doi:10.1103/PhysRevB.74.195310
13. Miotto, R., Srivastava, G.P.: Dissociative adsorption of PH_3 on the Si(001) surface. Phys. Rev. B **63**, 125321 (2001). doi:10.1103/PhysRevB.63.125321
14. Lin, D.-S., Ku, T.-S., Sheu, T.-J.: Thermal reactions of phosphine with Si(100): a combined photoemission and scanning-tunneling-microscopy study. Surf. Sci. **424**, 7 (1999). doi:10.1016/S0039-6028(98)00943-1
15. Perdew, J.P., Burke, K., Ernzerhof, M.: Generalized gradient approximation made simple. Phys. Rev. Lett. **77**, 3865 (1996). doi:10.1103/PhysRevLett.77.3865
16. Kresse, G., Hafner, J.: *Ab initio* molecular dynamics for liquid metals. Phys. Rev. B **47**, 558 (1993). doi:10.1103/PhysRevB.47.558
17. Kresse, G., Hafner, J.: *Ab initio* molecular dynamics simulation of the liquid-metal-amorphous-semiconductor transition in germanium. Phys. Rev. B **49**, 14251 (1994). doi:10.1103/PhysRevB.49.14251
18. Kresse, G., Furthmüller, J.: Efficient iterative schemes for *ab initio* total-energy calculations using a plane-wave basis set. Phys. Rev. B **54**, 11169 (1996). doi:10.1103/PhysRevB.54.11169
19. Kresse, G., Furthmüller, J.: Efficiency of ab-initio total energy calculations for metals and semiconductors using a plane-wave basis set. Comput. Mat. Sci. **6**, 15 (1996). doi:10.1016/0927-0256(96)00008-0
20. Kawai, H., Neucheva, O., Yap, T.L., Joachim, C., Saeys, M.: Electronic Characterization of a Single Dangling Bond on *n*- and *p*-type Si(001)-(2 × 1):H. Surf. Sci. Lett. **645**, L88–L92 (2016). doi:10.1016/j.susc.2015.11.001

Electronic Properties of a Single Dangling Bond and of Dangling Bond Wires on a Si(001):H Surface

Hiroyo Kawai, Olga Neucheva, Tiong Leh Yap, Christian Joachim and Mark Saeys

Abstract Single dangling bonds created on a Si(001):H surface are considered as potential fundamental building blocks for the fabrication of atom-scale electronic circuits. Therefore, it is critical to characterize the structural and electronic properties of single dangling bonds to begin to design and construct such devices. Using low-temperature scanning tunneling spectroscopy, imaging, density functional theory, and quantum transport calculations, we demonstrate the influence of doping, charging, and buckling on the electronic properties of a single dangling bond and a line of dangling bonds created along a dimer row.

Section 2: Reprinted with minor editorial changes from Surface Science Letters, **645**, Hiroyo Kawai, Olga Neucheva, Tiong Leh Yap, Christian Joachim, and Mark Saeys, "Electronic characterization of a single dangling bond on n- and p-type Si(001)-(2×1):H," L88-L92, 2016, with permission from Elsevier.

H. Kawai (✉) · O. Neucheva · T.L. Yap
Institute of Materials Research and Engineering, 2 Fusionopolis Way, Innovis, #08-03, Singapore 138634, Singapore
e-mail: kawaih@imre.a-star.edu.sg

T.L. Yap
GLOBALFOUNDRIES Singapore Pte Ltd., 60 Woodlands Industrial Park D Street 2, Singapore 738406, Singapore

T.L. Yap
Department of Physics, National University of Singapore, 2 Science Drive 3, Singapore 117542, Singapore

C. Joachim
GNS & MANA Satellite, CEMES-CNRS, 29 rue J. Marvig, 31055 Toulouse Cedex, France

C. Joachim
International Center for Materials Nanoarchitectronics (WPI-MANA), National Institute for Materials Science (NIMS), 1-1 Namiki, Tsukuba Ibaraki 305-0044, Japan

M. Saeys
Laboratory for Chemical Technology, Ghent University, Technologiepark 914, 9052 Ghent, Belgium

M. Kolmer and C. Joachim (eds.), *On-Surface Atomic Wires and Logic Gates*, Advances in Atom and Single Molecule Machines,
DOI 10.1007/978-3-319-51847-3_7

1 Introduction

The advancement in scanning tunneling microscopy (STM) technology has allowed the controllable removal of a single hydrogen atom from the passivated Si(001):H [1–10] and Ge(001):H surfaces [11, 12] to create a dangling bond (DB), thereby introducing an electronic state in the surface band gap. The single DB states are proposed as fundamental building blocks for the construction of atom-scale logic gates with a minimum number of quantum states [13, 14]. Thus, it is critical to fully understand the electronic properties of the single DB and the interaction between neighboring DBs. In this chapter, the electronic properties of a single DB on both n- and p-type Si(001):H surfaces were investigated using both experimental and theoretical approaches. The STM images and d*I*/d*V* spectra for the single DB were recorded, and the effect of bulk doping level on the images and electron tunneling transmission spectra through a single DB were explained using density functional theory (DFT) studies and quantum transport simulations. Subsequently, the electronic properties of a line of single DBs created on n-type Si(001):H were studied using similar approaches.

2 Single DB on a Si(001):H Surface

Single DBs were created on both n- and p-type Si(001):H surfaces using STM tip manipulation at low temperature, and the STM images and d*I*/d*V* spectra for the single DB were recorded (Fig. 1). The experiments were carried out in an ultrahigh vacuum (UHV) system with a base pressure below 5×10^{-10} mbar. The STM measurements were taken with an Omicron low-temperature (4.2 K) STM. See ref. [15] for the sample preparation procedures.

In Fig. 1a, b spectra measured for an STM tip positioned above the DB (blue curve) and above the hydrogen-passivated Si(001):H surface (red curve) are compared. For both n- and p-type samples, the d*I*/d*V* spectrum measured on the hydrogen-passivated Si(001):H surface is flat in the voltage range shown. For the n-type sample, the d*I*/d*V* spectrum measured above the DB shows a sharp peak at a bias around −1.6 V and a large background beyond −2.0 V, whereas for the p-type sample, the d*I*/d*V* spectrum measured above the DB shows a peak for a positive bias of about +1.3 V.

The STM images of a single DB on n- and p-type Si(001):H were compared to further illustrate the difference between them (Fig. 2a, c). A filled-state image of a DB on n-type Si(001):H at −2.1 V (Fig. 2a, −2.1 V) is round and bright due to the high conductance resulting from an increased coupling between Si bulk and the tip electronic states through the DB electronic states [16]. A DB appears darker at −1.5 V (Fig. 2a, −1.5 V), corresponding to the voltage range between −1.5 and 0 V in the d*I*/d*V* plot where no peaks are observed. Note that the bias voltage in d*I*/d*V* measurements and in the STM image measurements does not exactly match

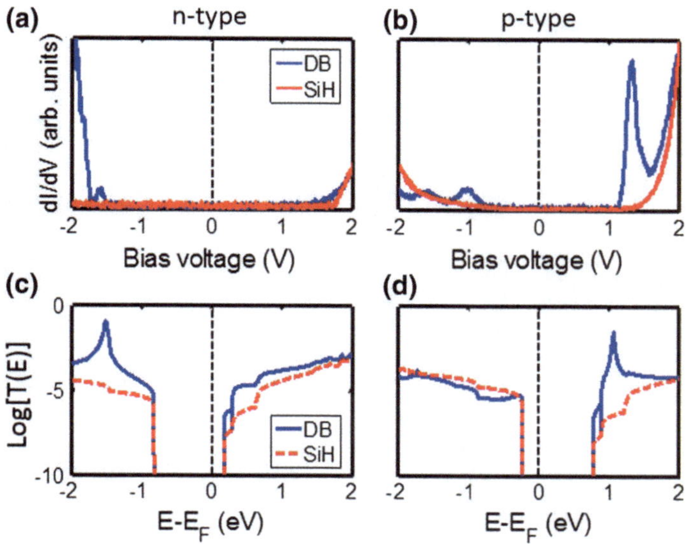

Fig. 1 dI/dV spectra for **a** n-type and **b** p-type Si(001):H surface measured for a tip above the DB (*blue curve*) and above the hydrogen-passivated surface (*red curve*). Transmission spectra of the STM tunneling junction calculated for a tip positioned above the DB (*blue curve*) and above the center of a hydrogen-passivated dimer (*dashed red curve*) on **c** n-type and **d** p-type Si(001):H

due to the different set point parameters and hence different amounts of tip-induced band bending (TIBB) for each measurement. The empty-state image of a DB on n-type Si at +1.5 V appears bright (Fig. 2a, +1.5 V), corresponding to the larger dI/dV at the conduction band edge compared to the hydrogen-passivated surface (Fig. 1a). At a higher bias of +2.1 V, a dark halo is found around the DB, except for a small protrusion at the position of the DB (Fig. 2a, +2.1 V). A dark halo is frequently observed for images of a single DB on an n-type Si(001):H at high positive bias [4, 7, 9, 17] and is caused by the charging of the DB state by the tip during the empty-state measurements. This effect induces screening in the area surrounding the DB [17].

For p-type Si(001):H, the empty-state images of the DB at +1.5 and +2.0 V appear bright (Fig. 2c), corresponding to the peaks observed at +1.3 and +2.0 V in the dI/dV spectrum (Fig. 1b), respectively. The image of a DB on a p-type substrate is significantly different from the image of a DB on an n-type substrate. The image of a DB on a p-type substrate has a distinct peanut-like shape, while it is round and symmetric for the n-type substrate. The filled-state image of a DB on p-type sample at low negative bias voltage appears darker (Fig. 2c, −1.2 V), but at larger negative bias, the DB starts to appear slightly brighter (Fig. 2c, −1.5 V). This change in brightness corresponds to the changes in the dI/dV spectra with bias voltage. Note that there is a small mismatch in the bias voltage due to the different set point parameters for STS and STM measurements, similar to the measurements for n-type DB. The filled-state image of a DB at larger negative bias shows the bright

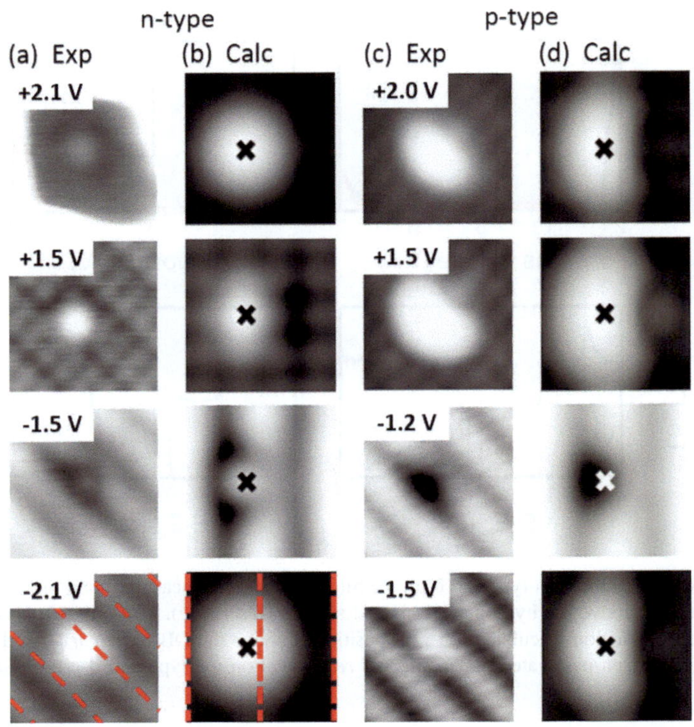

Fig. 2 Experimental and calculated STM images of a single DB on an **a**, **b** n-type and **c**, **d** p-type Si(001):H surface, respectively, obtained at different bias voltages. (experiment: n-type: I = 20 pA, p-type: I = 50 pA, 4 × 4 nm scan area; Calculated: 20 pA current, 1.5 × 1.5 nm scan area for both n-type and p-type, see the inset of Fig. 2a for the calculated area.) The DB position is indicated for the calculated images. The dimer with the DB is buckled by +7.6° and −10.2° to model n-type and p-type Si(100):H, respectively (see Fig. 2). *Red dashed lines* on the experimental and calculated images of the DB on n-type substrate at −2.1 V indicate the centers of the dimer rows

protrusion that extends along the dimer rows, which is again different from the round and symmetric filled-state image for the n-type DB.

In order to identify the states observed in the d*I*/d*V* spectra and to rationalize the change in contrast to the STM images, the electronic properties of a single DB on both n- and p-type Si(001):H were investigated using DFT and quantum transport simulations. The Si(001):H surface was modeled as a 13-layer slab with a 4 × 4 supercell with periodic boundary conditions. The top and bottom surfaces were passivated with hydrogen, except for the single DB to be investigated. A 12 Å vacuum gap is introduced to decouple repeated slabs in the periodic calculations. The surface structures were optimized using DFT with the Perdew–Burke–Ernzerhof (PBE) functional [18], as implemented in the Vienna *ab initio* simulation package (VASP) [19–22]. Electrons were added to or removed from the unit cell to model doping in n-type and p-type Si, respectively, and the band structures and

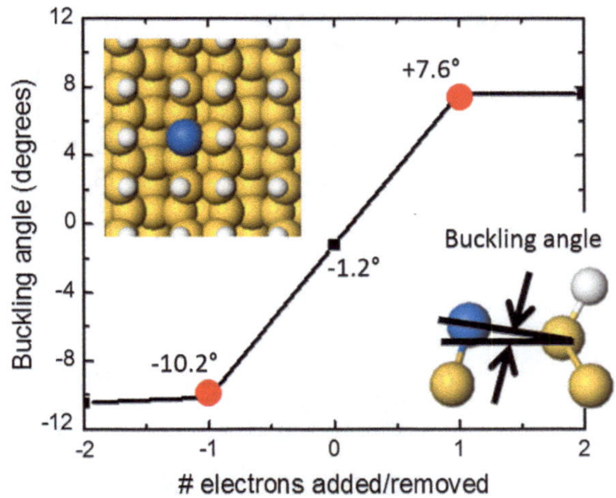

Fig. 3 Change in the buckling angle of a Si dimer as a function of charging of the DB for an isolated DB in a 4 × 4 unit cell as shown in the inset. The geometries used to model a single DB on n-type and p-type Si(001):H are shown by the *red dots*. The buckling angle is indicated in the side view of the dimer

charge densities of the n- and p-type Si slabs were calculated using DFT. STM images and transmission spectra, $T(E)$, were calculated using the surface Green function matching (SGFM) method [23] with an extended Hückel molecular orbital (EHMO) Hamiltonian. The STM model fully accounts for quantum tunneling from the STM tip to the Si(001):H substrate. The TIBB was incorporated by shifting the energies of the surface states relative to the bulk bands, and the amount of TIBB was approximated from the experimentally measured dI/dV and STM images and estimates from electrostatic simulations [9, 24].

The geometry of the dimer with a single DB was optimized for a range of charge states (Fig. 3). The calculations show that the buckling angle is very sensitive to the charge on the single DB Si atom. When the DB Si atom is close to neutral, the dimer is nearly unbuckled. Increasing the electron density, which corresponds to n-type doping, gradually increases the buckling angle until it saturates at around 8°. Decreasing the electron density, which corresponds to p-type doping, gradually reduces the buckling angle until it saturates at around −10°, with the DB Si atom being lower than the H-passivated Si atom in the dimer. Saturation occurs when the charge on the DB Si atom reaches about +1 or −1, respectively. The change in the buckling angle can be explained by electron pair repulsion [25].

As the DB Si atom becomes negatively charged (n-type doping), the hybridization evolves toward sp^3 to minimize repulsion between the electron pairs. Removing electron density from the DB Si atom (p-type doping) changes the hybridization toward sp^2, for the same reason. The saturated geometries (indicated by the red dots in Fig. 3) are used to model the structure of a single DB on n-type and p-type Si(001):H. However, the actual buckling will depend on the doping level and the bias voltage.

The band structures for the n- and p-type Si(001):H with a single DB are shown in Fig. 4a, b. For both substrates, removal of a hydrogen atom from a passivated surface introduces a DB state in the surface band gap, close to the surface valence

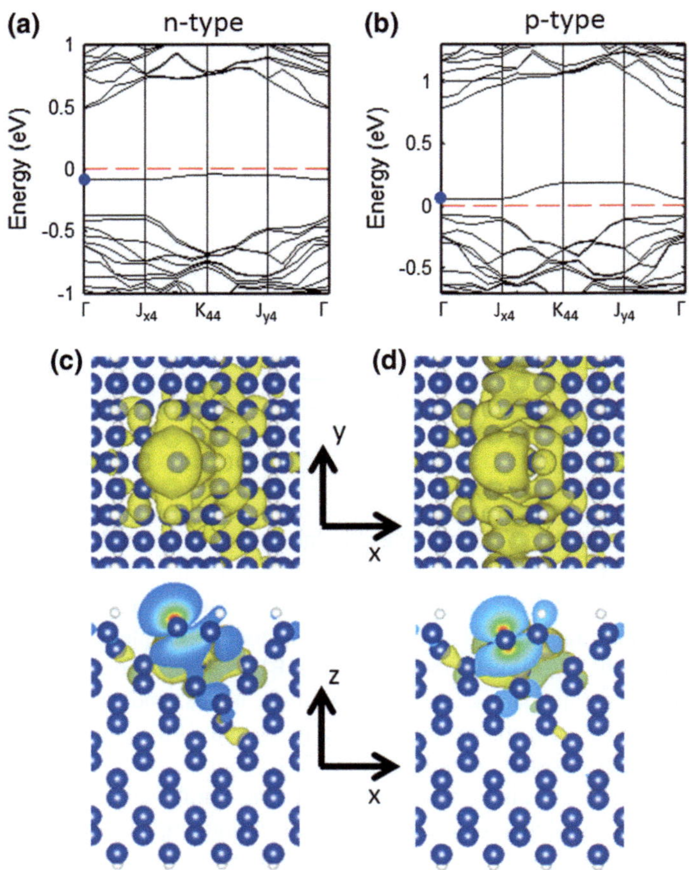

Fig. 4 **a**, **b** DFT surface band structure of Si(001):H with a single DB for n-type and p-type models, respectively. The band structures are aligned at the valence band edge. **c**, **d** The *top* and *side* views of the atomic structure and the DB state-associated electron probability density at Γ for n- and p-type Si(001):H, respectively, with an isosurface level of 7×10^{-4} e/Å^3. Note the different hybridization and expansion of the electron density

band edge. However, the DB state on p-type Si extends significantly along the dimer rows, as indicated by the dispersion in the band structure. This is also evident from the top view of the corresponding electron probability density plot (Fig. 4c, d). This difference is caused by the difference in the hybridization of the Si DB atom for both substrates. On n-type substrates, the DB state is nearly sp^3 hybridized, highly directional and coupled rather weakly with the bulk states. On p-type substrates, the DB Si atom is buckled down and is well coupled to the states along the dimer rows. It is expected that the very distinct shapes of the DB-associated electron probability density result in very different STM images.

Transmission spectra $T(E)$ of the STM junction were calculated for n- and p-type Si(001):H with the STM tip positioned above the DB (blue curve) and above the center of the hydrogen-passivated dimer [dashed red curve] (Fig. 1c, d). STM images were also calculated for n- and p-type Si(001):H (Fig. 2b, d, respectively) and were compared with the experimentally measured STM images. For n-type Si(001):H, TIBB shifts the DB state down by about 1 eV for a bias of −2.1 V, resulting in an intense resonance peak at −1.5 eV in the *T*(*E*) spectrum. No resonance peak is observed for positive bias voltages up to 2 V, but the STM junction conductance above the DB is higher than above the passivated dimer due to the upward buckling of the DB Si atom for n-type Si. The STM image calculated for a bias voltage of −2.1 V shows a symmetric and bright protrusion (Fig. 2b, −2.1 V) similar to the experimental STM image (Fig. 2a, −2.1 V), corresponding to the DB resonance at −1.5 eV coupled strongly to the bulk states. Note that in our *T*(*E*) and image calculations, the DB state is over-coupled to the bulk states compared to the experimental observations due to the approximations made for TIBB. In reality, the thickness of the charge accumulation/depletion layer near the surface is of the order of tens of nm [9], whereas in our calculations, this surface layer is less than 1 nm. The shape of the STM image reflects the directional sp^3-character of the DB state in Fig. 4c. For a smaller bias voltage of −1.5 V and correspondingly smaller TIBB, the calculated image shows a depression around the DB (Fig. 2b), similar to the experimental image. Because of the small TIBB for this bias voltage, the DB state falls in the surface band gap and the *T*(*E*) above the DB is smaller than the *T*(*E*) above a passivated dimer. Empty-state images for n-type Si show a bright protrusion at the position of the DB (Fig. 2b, +1.5 V), even though the DB state lies in the band gap for these bias voltages. The calculations suggest that the higher *T*(*E*) is caused by the enhanced coupling of the tip state with surface states near the conduction band edge through the directional sp^3-like DB state. For a larger positive bias, the dynamic charging of the DB state becomes significant, a dynamic effect which is not included in our quantum transport simulations.

The *T*(*E*) spectra and calculated STM images of the DB on a p-type substrate are significantly different from the results obtained for n-type substrate. TIBB shifts the DB state upwards by about 1 eV for positive bias voltages of about 1.5–2.0 V, enough to shift the DB-derived state into the conduction band and resulting in a resonance in the conduction band at around +1.1 eV (Fig. 1d), similar to the d*I*/d*V* curve in Fig. 1b. Note that similar to the n-type case, the over-coupling of the DB states to the bulk states in the calculations causes this peak to be higher than the *T*(*E*) at higher voltages. The empty-state images of the DB state at +1.5 and +2.0 V are clearly expanded along the dimer row (Fig. 2d) with distinct peanut-like shape, in agreement with the experimental empty-state images (Fig. 2c). The expansion of the DB image results from the strong coupling of the p_z-like DB state on the p-type Si substrate with surface states at the conduction band edge, as illustrated in Fig. 4d. The significant lateral expansion is hence a signature of the different hybridization and buckling of a DB Si atom on p-type Si(001):H. The filled-state images at −1.2 V show a depression at the position of the DB corresponding to the lower *T*(*E*). At larger negative bias, TIBB shifts the DB state into the valence band also on p-type Si(001):H, resulting in a bright filled-state image. Again, the protrusion associated with the DB is expanded along the dimer rows.

Based on the STM studies combined with the DFT analysis and quantum transport calculations, doping is shown to have a major influence on the geometry and electronic properties of the DB state. On p-type Si, the STM images show that the DB states couple stronger to the surface states resulting in a significant expansion along the dimer rows. In contrast, on n-type Si(001):H, a highly symmetric image is obtained. The next question is how the single DBs interact when a line of DBs is created on the Si(001):H surface.

3 DB Lines Created on a Si(001):H Surface

In order to investigate interactions between neighboring DBs on a Si(001):H surface, short lines of one to five single DBs were created along a dimer row, as illustrated in Fig. 5a. Along the lines, the DBs are separated by a H-saturated Si dimer. The short DB lines were analyzed using the experimental and modeling approaches described in Sect. 2. Only DB lines created on n-type Si(001):H will be discussed. The DBs in the lines are hence expected to be (partially) negatively charged and buckled up, as shown in the previous section.

Figure 5b, c show the filled-state STM images and the d*I*/d*V* spectra for wires ranging from one to five DBs. For each wire, d*I*/d*V* spectra were obtained for a tip positioned above each of the DBs. The numbers in the d*I*/d*V* spectra indicate the DB above which the spectrum was acquired.

The image for a ***single DB*** appears similar to the image in Fig. 2a (n-type, −2.1 V), and also the d*I*/d*V* spectrum shows a peak at negative bias (−2.1 V) and no peak for the positive bias range. The position of the peak in Fig. 5c is slightly different from the one in Fig. 1a due to the different set point values and thus the different extent of TIBB. The image for a wire of ***two DBs*** appears asymmetric with one brighter DB, which may be due to a slight difference in buckling for the 2 DBs. The d*I*/d*V* spectra obtained above each of the DBs show peaks in the negative bias range, but with different intensity and width. The peak obtained when the tip is positioned above the brighter DB (DB1) is intense and sharp, while the peak obtained above DB2 is less intense and broader. However, both peaks are centered on the same energy (−2.1 V), and no peak is observed for the positive bias range. The filled-state image for a wire of ***three DBs*** also appears asymmetric and DB3 has likely shifted to the other side of the dimer due to a shift of the H atom from one side of the dimer to the other during the measurements. The d*I*/d*V* peak for DB1 is intense and sharp, while the d*I*/d*V* peaks for DB2 and DB3 are less intense and broader, similar to the peak for DB2 in the line of two DBs. Again, all the peaks are centered around −2.1 V. Interestingly, the d*I*/d*V* spectrum for the central DB (DB2) shows a small peak for the most positive bias range (+2.1 V) while no peaks are recorded for positive bias for the edge DBs.

The STM image for a wire of ***four DBs*** appears quite symmetric, and the edge DBs are slightly brighter than the inner DBs. The d*I*/d*V* spectrum for the edge DBs again displays a sharp peak around −2.0 V, while the peaks for the inner DBs appear

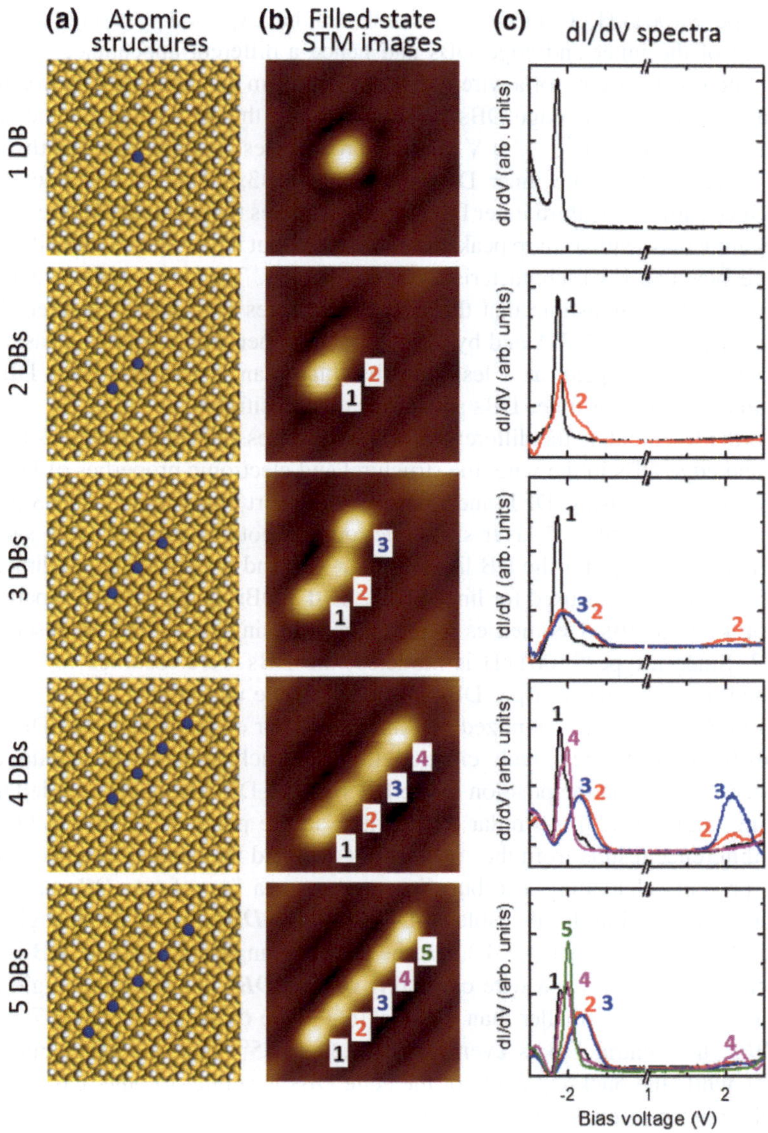

Fig. 5 **a** Atomic structures, **b** filled-state STM images at −2.0 V, and **c** d*I*/d*V* spectra for DB lines of one to five DBs. The d*I*/d*V* spectra were obtained with the STM tip positioned above each individual DB. The DBs are labeled in the STM images and correspondingly in the d*I*/d*V* spectra. (STM images: I = 20 pA, scan area 4 × 4 nm)

less intense, broader, and slightly shifted toward a lower bias (−1.6 V). For the inner DBs, a small peak is also recorded for the positive bias range (+2.1 V). Both the STM image and d*I*/d*V* spectra suggest that the inner DBs in the wire are electronically and/or structurally different from the edge DBs. Since DBs are likely charged on n-type Si(001):H, Coulomb repulsion can be expected, causing a different charging of the inner and edge DBs and hence a different appearance. The STM image and d*I*/d*V* spectra for a wire of ***five DBs*** confirm the distinction between edge and inner DBs. For the edge DBs (DB1 and DB5), the d*I*/d*V* displays the characteristic sharp peak around −2.0 V, while a broader, less intense and slightly shifted peak is recorded for the inner DBs (DB2 and DB3, −1.6 V). Interestingly, the d*I*/d*V* spectrum for the third inner DB (DB4) resembles the spectrum for the edge DB for negative bias, with a sharp peak around −2.0 V, but it also shows a small peak for positive bias (+2.1 V), characteristic for inner DBs. The d*I*/d*V* spectra and STM images clearly demonstrate that the DBs in the wires are interacting even though they are separated by 7.7 A and by a H-saturated dimer. d*I*/d*V* spectra measured for inner DBs show a peak at a less negative bias than spectra for edge DBs and sometimes spectra for inner DBs show a peak at positive bias.

In order to elucidate the differences in STM images and d*I*/d*V* peak positions for inner and edge DBs in the wire, the structural and electronic properties of DB wires were investigated using DFT and quantum transport simulations. The Si(001):H surface was modeled as 5-layer slab. The top and bottom surfaces were saturated with H atoms, except for the DB lines of 1, 2, 3, 4, and 5 DBs along the dimer row. A 2×8 supercell was used for lines of 1, 2, and 3 DBs, and a 2×12 supercell for lines of 4 and 5 DBs. Geometries were optimized using DFT-PBE, as described in Sect. 2. Since n-type Si(001):H is modeled, the DBs were assumed to be charge saturated, i.e., one electron per DB was added to the unit cell.

Figure 6a shows the optimized buckling angle for each dimer in the DB. Since the DB Si atoms are negatively charged, they are buckled up. The buckling angle, however, depends on the position of the DB within a DB wire. For a single DB, the buckling angle (+7.2°) is similar to the value in the previous section (+7.6°); the small difference results from the smaller supercell and the smaller number of layers in the present calculation. The buckling angles for a line of ***two DBs*** (+7.3°) are similar to the buckling for the isolated DB. For ***three DBs***, the buckling angle of the edge DBs is again similar (+7.3°), but the buckling angle for the inner DB is much smaller (+5.6°). Also for a wire consisting of ***four DBs***, the buckling angle of the inner DBs (+6.1°) is smaller than the buckling angle of the edge DBs (+7.3°). For ***five DBs***, the central DB is even less bucked (+4.5°) than the other inner DBs (+6.0°), while the buckling angle of the edge DBs is similar to other cases (+7.3°). The differences in buckling angle suggest difference in charging. Coulomb repulsion reduces the charge on the inner DBs and the small reduction in electron density reduces the buckling angles, as shown in Fig. 3. This shift in electrons does not increase the buckling of the edge DBs, as the DBs are saturated, as described in the previous section (see Fig. 3).

Since the buckling angle of edge DBs differs from the buckling angle of inner DBs, the corresponding DB state is expected to be shifted away from the valence band edge. The EHMO band structures for a Si(001):H surface with DB lines of 1–5 DBs are shown in Fig. 7. The DBs indeed introduce DB states within the surface gap, and by calculating the projected density of states (not shown), the DB that contributes most to each state in the gap can be determined and is labeled.

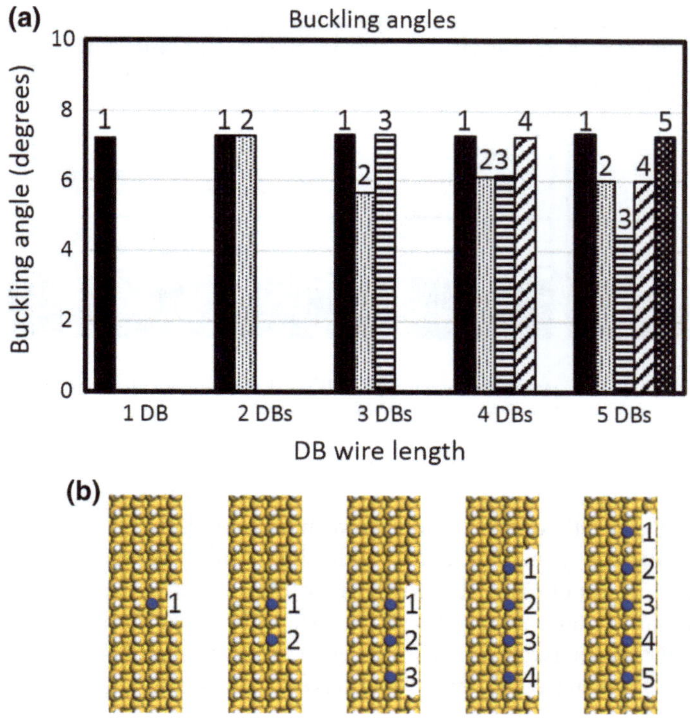

Fig. 6 **a** DB buckling angles for DB wires of different lengths. Each number corresponds to the label in **b**. **b** Atomic structures of the Si(001):H surface with wires of 1–5 DBs along the dimer row

Figure 7a illustrates that edge DB states are about 0.4 eV below the center of the surface band gap (Fig. 7b–e), similar to the energy position of a single DB state (Fig. 7a). However, inner DB states are about 0.3–0.4 eV higher in energy than edge DB states, and further from the valence band edge. The state corresponding to the central DB of a 5 DB line (Fig. 7e, "3") is shifted even slightly further from the conduction band edge. The results show that the smaller the buckling angle, the higher the energy position of the DB states, as expected from the increased p-character of the DB state.

Using these EHMO band structures for the STM model, the $T(E)$ spectra and STM images were calculated. The $T(E)$ spectra and the corresponding calculated STM images are shown in Fig. 8a, b, respectively. The $T(E)$ spectra were calculated with the tip positioned above each individual DB, as in the experiments. Note that due to symmetry, the two edge DBs have the same calculated $T(E)$, and also the two inner DBs, such as 2 and 4 in the 5 DB wire, have the same $T(E)$. As expected from the band structures, the $T(E)$ spectra calculated with the tip positioned above the inner DBs show peaks at higher energy (i.e., closer to the gap) compared to the $T(E)$ calculated with the tip positioned above the edge DBs. For the line of 5 DBs,

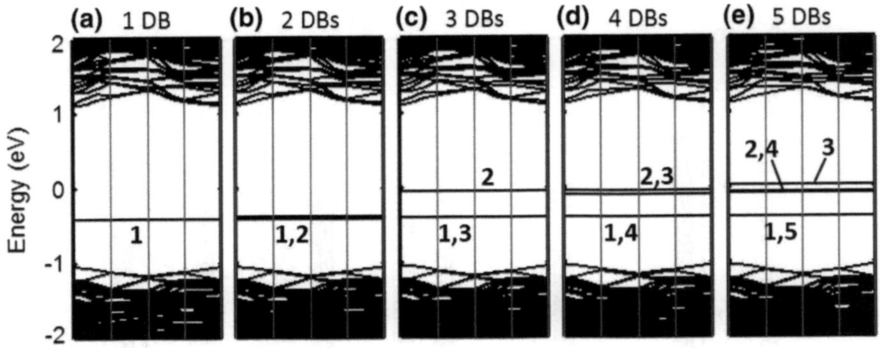

Fig. 7 **a–e** EHMO band structures for a n-type Si(001):H surface with DB lines of 1–5 DBs along the dimer row. See Fig. 6b for the supercell structures and illustration of the labels

the *T*(*E*) peak appears at even higher energy when the tip is positioned above the central DB ("3"), although the difference with the inner DBs ("2" and "4") is small. These simulations generally agree with the trends observed experimentally. *T*(*E*) spectra were not calculated for positive bias, but it is expected that upward TIBB will shift states associated with the inner DB closer to the conduction band edge than the states associated with the edge DBs and that therefore (part of) the *T*(*E*) peak for the inner DBs may be observed at the CBE, as seen in the experimental d*I*/d*V* curves (Fig. 5c).

The calculated STM images for DB wires of various length are shown in Fig. 8b. The images were calculated for different bias voltages to show how the contrast changes with bias. For simplicity, the same amount of TIBB was assumed for all voltages. At −2.1 V, the DBs for wires with one DB and with two DBs have a similar contrast. This is expected since the buckling angles as well as the calculated peak positions are similar for these DBs. For wires of 3 and 4 DBs, the brightness of the edge DBs is similar to the brightness for 1 and 2 DBs, but the inner DBs appear less bright at −2.1 V due to their smaller buckling (i.e., the inner DBs are lower in height). The central DB in 5 DB wire is even less bright due to the smaller buckling angle and the shift in peak position compared to the other two inner DBs. The STM images calculated at −1.4 V, where the *T*(*E*) peak of edge DB is observed, are similar to those calculated at −2.1 V, except that the contrast between the edge DBs and the inner DBs is smaller. The STM images calculated at −1.3 V, however, are significantly different. In these cases, the inner DBs appear brighter than the edge DBs because the edge DB states fall outside the voltage range and thus do not contribute to the STM image at this smaller bias. The experimentally observed contrast in Fig. 5b is closer to the calculated images at −1.4 V. The difference in brightness between the inner and the edge DBs in the calculated images at −2.1 V may be exaggerated due to the over-coupling of the DB states with the bulk states, as discussed in the previous section. Another possible reason for the exaggerated contrast is that the buckling angle in the calculations may be larger than the actual buckling angle. It was assumed that for n-type Si(001):H, all DBs are saturated, but

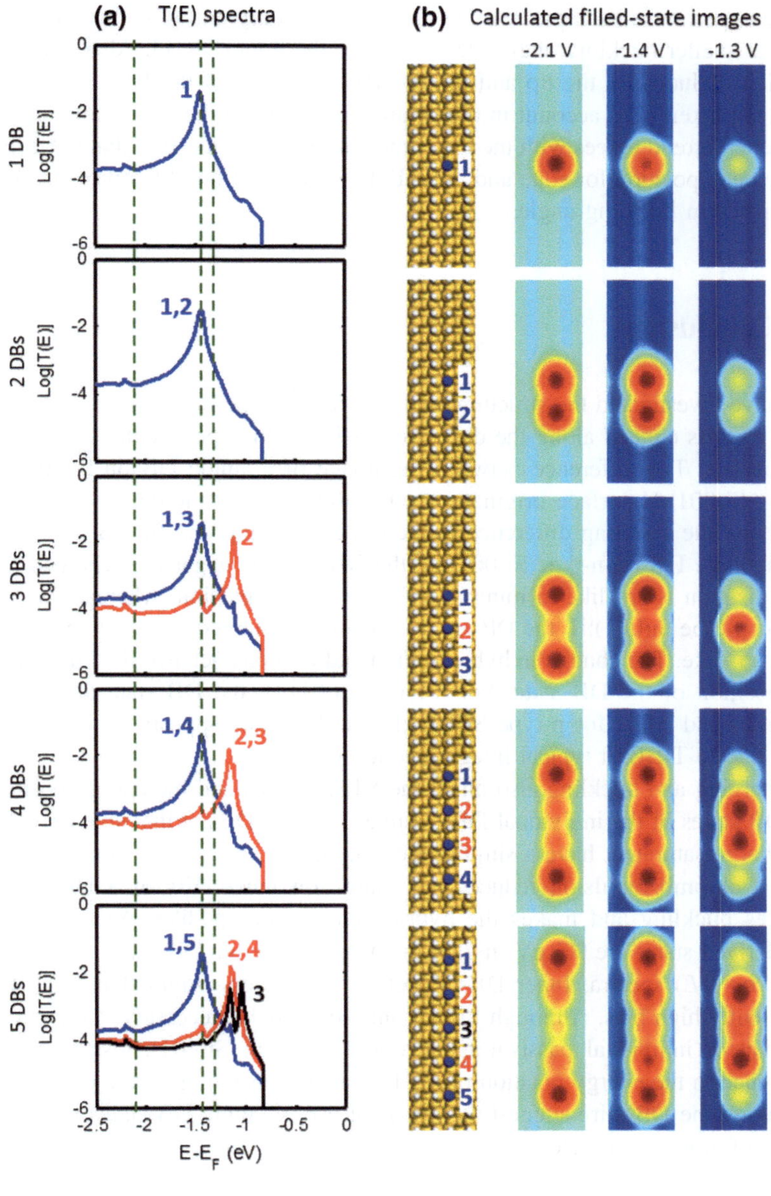

Fig. 8 **a** *T*(*E*) spectra calculated with the STM tip positioned above each individual DB and **b** calculated STM images at −2.1, −1.4, and −1.3 V for lines of one to five DBs along the dimer row. The *T*(*E*) peaks are labeled corresponding to the DB position indicated in the atomic structures

in the experimental sample, the DBs are possibly only partially charged (and thus have a smaller buckling angle) depending on the doping level. Also, the effect of the field induced by the tip and the dynamic charging of the DB during imaging were not taken in to account in the simulations. Despite the approximations made, the general trend agrees with the experimental results, suggesting that the difference in the peak position for edge and inner DBs originates from differences in charging and hence in buckling angle.

4 Conclusions

We have investigated the structural and electronic properties of a single DB and of lines of DBs created along the dimer row using both experimental and theoretical approaches. The difference between the image of a single DB on n-type and on p-type Si(001):H surface originates from a difference in the charge of the DB Si atom and the resulting difference in the dimer buckling and Si DB hybridization. For a single DB on n-type Si(001):H, the DB atom is negatively charged and the hybridization is sp^3-like to minimize electron pair repulsion, whereas for a single DB on p-type Si(001):H, the DB atom is positively charged and the hybridization is more sp^2-like. The change in hybridization and in *p*-character of the DB state shifts the position of the DB state. For n-type Si(001):H, the DB state appears in the valence band while for p-type Si(001):H, the DB state appears in the conduction band due to TIBB at typical imaging conditions.

Charging and buckling also affect the STM images and the energy positions of the DB states of the individual DBs in lines of DBs. DBs at the edge of a DB line are charge saturated, have a similar buckling angle, and appear similar to isolated DBs. Coulomb repulsion reduces the charge on inner DBs of a DB line. This reduces buckling and makes the hybridization more sp^2-like. As a result, inner DB-derived states are 0.3 eV higher in energy than edge DB states, as seen in the dI/dV or $T(E)$ spectra. Inner DBs therefore appear less bright than edge DBs at sufficiently high bias. Although further analysis may be necessary, the controllable removal of individual passivating H atoms from the Si(001):H surface and the difference in the energy position of the DB states depending on the position of the DB along the DB wire suggest the potential use of single DB states for the fabrication of atomic-scale electronic devices.

Acknowledgements We acknowledge the Agency of Science, Technology, and Research (A*STAR) for funding provided through the Visiting Investigatorship Programme Atom Technology Project 1021100972, and through the AtMol Integrated Project Contract Number 270028 from the European Commission. We also acknowledge the A*STAR Computational Resource Centre (A*CRC) for computational resources and support.

References

1. Shen, T.-C., Wang, C., Abeln, G.C., Tucker, J.R., Lyding, J.W., Avouris, Ph, Walkup, R.E.: Atomic-scale desorption through electronic and vibrational excitation mechanisms. Science **268**, 1590 (1995). doi:10.1126/science.268.5217.1590
2. Lyding, J.W., Shen, T.-C., Hubacek, J.S., Tucker, J.R., Abeln, G.C.: Nanoscale patterning and oxidation of H-passivated Si(100)-2×1 surfaces with an ultrahigh vacuum scanning tunneling microscope. Appl. Phys. Lett. **64**, 2010 (1994). doi:10.1063/1.111722
3. Bianco, F., Bowler, D.R., Owen, J.H.G., Köster, S.A., Longobardi, M., Renner, C.: Scalable patterning of one-dimensional dangling bond rows on hydrogenated Si(001). ACS Nano **7**, 4422 (2013). doi:10.1021/nn4010236
4. Schofield, S.R., Studer, P., Hirjibehedin, C.F., Curson, N.J., Aeppli, G., Bowler, D.R.: Quantum engineering at the silicon surface using dangling bonds. Nat. Commun. **4**, 1649 (2013). doi:10.1038/ncomms2679
5. Kolmer, M., Godlewski, S., Zuzak, R., Wojtaszek, M., Rauer, C., Thuaire, A., Hartmann, J.-M., Moriceau, H., Joachim, C., Szymonski, M.: Atomic scale fabrication of dangling bond structures on hydrogen passivated Si(001) wafers processed and nanopackaged in a clean room environment. Appl. Surf. Sci. **288**, 83 (2014). doi:10.1016/j.apsusc.2013.09.124
6. Soukiassian, L., Mayne, A.J., Carbone, M., Dujardin, G.: Atomic wire fabrication by STM induced hydrogen desorption. Surf. Sci. **528**, 121–126 (2003). doi:10.1016/S0039-6028(02) 02620-1
7. Bellec, A., Riedel, D., Dujardin, G., Boudrioua, O., Chaput, L., Stauffer, L., Sonnet, P.: Electronic properties of the *n*-doped hydrogenated silicon (100) surface and dehydrogenated structures at 5 K. Phys. Rev. B **80**, 245434 (2009). doi:10.1103/PhysRevB.80.245434
8. Naydenov, B., Boland, J.J.: Engineering the electronic structure of surface dangling bond nanowires of different size and dimensionality. Nanotechnology **24**, 275202 (2013). doi:10.1088/0957-4484/24/27/275202
9. Taucer, M., Livadaru, L., Piva, P.G., Achal, R., Labidi, H., Pitters, J.L., Wolkow, R.A.: Single-electron dynamics of an atomic silicon quantum dot on the H-Si(100)-(2 × 1) surface. Phys. Rev. Lett. **112**, 256801 (2014). doi:10.1103/PhysRevLett.112.256801
10. Bellec, A., Chaput, L., Dujardin, G., Riedel, D., Stauffer, L., Sonnet, P.: Reversible charge storage in a single silicon atom. Phys. Rev. B **88**, 241406(R) (2013). doi:10.1103/PhysRevB. 88.241406
11. Kolmer, M., Godlewski, S., Kawai, H., Such, B., Krok, F., Saeys, M., Joachim, C., Szymonski, M.: Electronic properties of STM-constructed dangling-bond dimer lines on a Ge(001)-(2×1): H surface. Phys. Rev. B **86**, 125307 (2012). doi:10.1103/PhysRevB.86. 125307
12. Kolmer, M., Godlewski, S., Lis, J., Such, B., Kantorovich, L., Szymonski, M.: Construction of atomic-scale logic gates on a surface of hydrogen passivated germanium. Microelectron. Eng. **109**, 262 (2013). doi:10.1016/j.mee.2013.03.061
13. Kawai, H., Ample, F., Wang, Q., Yeo, Y.K., Saeys, M., Joachim, C.: Dangling-bond logic gates on a Si(100)-(2×1)-H surface. J. Phys.: Condens. Matter **24**, 095011 (2012). doi:10.1088/0953-8984/24/9/095011
14. Ample, F., Duchemin, I., Hliwa, M., Joachim, C.: Single OR molecule and OR atomic circuit logic gates interconnected on a Si(100)H surface. J. Phys.: Condens. Matter **23**, 125303 (2011). doi:10.1088/0953-8984/23/12/125303
15. Kawai, H., Neucheva, O., Yap, T.L., Joachim, C., Saeys, M.: Electronic characterization of a single dangling bond on *n*- and *p*-type Si(001)-(2×1):H. Surf. Sci. Lett. **645**, L88–L92 (2016). doi:10.1016/j.susc.2015.11.001
16. Yap, T.L., Kawai, H., Neucheva, O.A., Wee, A.T.S., Troadec, C., Saeys, M., Joachim, C.: Si(100)-2×1-H dimer rows contrast inversion in low-temperature scanning tunneling microscope images. Surf. Sci. **632**, L13–L17 (2015). doi:10.1016/j.susc.2014.10.016

17. Livadaru, L., Pitters, J., Taucer, M., Wolkow, R.A.: Theory of nonequilibrium single-electron dynamics in STM imaging of dangling bonds on a hydrogenated silicon surface. Phys. Rev. B **84**, 205416 (2011). doi:10.1103/PhysRevB.84.205416
18. Perdew, J.P., Burke, K., Ernzerhof, M.: Generalized gradient approximation made simple. Phys. Rev. Lett. **77**, 3865 (1996). doi:10.1103/PhysRevLett.77.3865
19. Kresse, G., Hafner, J.: *Ab initio* molecular dynamics for liquid metals. Phys. Rev. B **47**, 558 (1993). doi:10.1103/PhysRevB.47.558
20. Kresse, G., Hafner, J.: *Ab initio* molecular-dynamics simulation of the liquid-metal-amorphous-semiconductor transition in germanium. Phys. Rev. B **49**, 14251 (1994). doi:10.1103/PhysRevB.49.14251
21. Kresse, G., Furthmüller, J.: Efficient iterative schemes for *ab initio* total-energy calculations using a plane-wave basis set. Phys. Rev. B **54**, 11169 (1996). doi:10.1103/PhysRevB.54.11169
22. Kresse, G., Furthmüller, J.: Efficiency of ab-initio total energy calculations for metals and semiconductors using a plane-wave basis set. Comput. Mater. Sci. **6**, 15 (1996). doi:10.1016/0927-0256(96)00008-0
23. Cerdá, J., Van Hove, M.A., Sautet, P., Salmeron, M.: Efficient method for the simulation of STM images. I. generalized green-function formalism. Phys. Rev. B **56**, 15885 (1997). doi:10.1103/PhysRevB.56.15885
24. Feenstra, R.M.: Electrostatic potential for a hyperbolic probe tip near a semiconductor. J. Vac. Sci. Technol., B **21**, 2080 (2003). doi:10.1116/1.1606466
25. Gillespie, R.J., Hargittai, I.: The VSEPR Model of Molecular Geometry (Reprint edition), p. 7. Dover Publications, New York (1991)

Quantum Hamiltonian Computing (QHC) Logic Gates

Hiroyo Kawai, Olga Neucheva, Ghassen Dridi, Mark Saeys and Christian Joachim

Abstract The quantum graph of a two-input–one-output QHC Boolean logic gate was used to design a very simple dangling bond QHC logic gate. On an Si(100):H surface and in the vertical single atom manipulation mode, an LT-UHV-STM was used for atom-by-atom construction of this QHC logic gate using only three dangling bonds. As expected, the experimental measurement shows a NOR gate at a 1.3-V positive bias voltage and an OR gate at −1.8 V. The same Boolean logical functions were also obtained in our quantum tunneling calculations, where the geometry and the electronic properties of the Si surface were determined using DFT. The surface Green function matching (SGFM) method was used to calculate the d*I*/d*V* spectra in a vertical STM measurement configuration and the tunneling current intensity through the DB gate in a planar two metallic nanopads configuration, confirming the NOR and the OR function of the QHC gate.

H. Kawai (✉) · O. Neucheva
Institute of Materials Research and Engineering, 2 Fusionopolis Way, Innovis, #08-03, Singapore 138634, Singapore
e-mail: kawaih@imre.a-star.edu.sg

G. Dridi · C. Joachim
GNS & MANA Satellite, CEMES-CNRS, 29 Rue J. Marvig, 31055 Toulouse Cedex, France

M. Saeys
Laboratory for Chemical Technology, Ghent University, Technologiepark 914, 9052 Ghent, Belgium

C. Joachim
International Center for Materials Nanoarchitectronics (WPI-MANA), National Institute for Materials Science (NIMS), 1-1 Namiki, Tsukuba, Ibaraki 305-0044, Japan

M. Kolmer and C. Joachim (eds.), *On-Surface Atomic Wires and Logic Gates*, Advances in Atom and Single Molecule Machines,
DOI 10.1007/978-3-319-51847-3_8

1 Introduction

The quantum Hamiltonian computing (QHC)-based Boolean logic approach is based on the control of the intrinsic time-dependent evolution of a well-designed quantum system by its logical inputs. Considering a quantum system (e.g., the π electrons of a molecule or, as in this chapter, a surface DB network) prepared in a non-stationary initial state $|\psi(0)\rangle$, its intrinsic response is a spontaneous time evolution described by the time-dependent state vector $|\psi(t)\rangle = e^{-iHt/\hbar}|\psi(0)\rangle$ where H is the Hamiltonian of the quantum system. Any arithmetic or logic operations performed by this quantum system can be interpreted as a $|\psi(t)\rangle\langle\psi(t)|$ trajectory control on its quantum state space. The logical inputs controlling this trajectory can be carried by the initial preparation $|\psi(0)\rangle$ or in the Hamiltonian H. These operations have normally to be performed before the decoherence of the initial $|\psi(0)\rangle$ wave packet and before the initial preparation relaxes to the ground state of the quantum system. Notice that one solution to benefit from relaxation and decoherence is to perform a large number of $|\psi(0)\rangle$ "preparation and decoherence (relaxation)" cycles that can be regularly or randomly spaced in time and to measure the average result of the Boolean operation at the end of those cycles as explored in this chapter using a tunneling junction either with an STM tip apex (see Sect. 4) or in a planar configuration using metallic nanopads (see Sect. 5).

The electronic structure of the gate, i.e., the H matrix elements, depends on the practical way chosen to encode the logical inputs and to measure the logical outputs of the gate. Formally, a $\{\alpha_1, \ldots, \alpha_i, \ldots, \alpha_k\}$ digital input word can be encoded either on $|\psi(0)\rangle$ or on some of the H matrix elements of the gate. When $|\psi(0)\rangle$ is encoding the logical inputs, then H is fixed. When $\{\alpha_1, \ldots, \alpha_i, \ldots, \alpha_k\}$ is encoded on the H matrix elements, then $|\psi(0)\rangle$ is fixed. A given $\{\beta_1, \ldots, \beta_i, \ldots, \beta_{k'}\}$ output word is associated with a given target state $|\phi\rangle$, which must be a pointer state for the time dependant Heisenberg–Rabi evolution. Specific characteristics of the $|\langle\phi|\psi(t)\rangle|^2$ population such as its maximum amplitude or its secular oscillation frequency can be used to define an appropriate output measurement strategy.

A first class of logic gates is obtained by encoding the logical input in different initial $|\psi(0)\rangle$ states and by measuring the logical output on the population amplitude of carefully selected target states. The truth table of those gates is defined by a linear and unitary transformation $B\ |\alpha_1, \ldots, \alpha_i, \ldots, \alpha_k\rangle = |\beta_1, \ldots, \beta_i, \ldots, \beta_{k'}\rangle$. The B operator is constructed for the result of the Boolean calculations to be measurable exactly at a series of time t_n and the Hamiltonian reads: $H = (i\hbar/t_n)\ \mathrm{Log}(B)$ because in such a qubit design, the $|\psi(t)\rangle\langle\psi(t)|$ quantum trajectory must be fully periodic.

In this chapter, we are following a second type of approach by encoding the $\{\alpha_1, \ldots, \alpha_i, \ldots, \alpha_k\}$ digital input words on the Hamiltonian, i.e., $H = H(\alpha_1, \ldots, \alpha_i, \ldots, \alpha_k)$ while the $\{\beta_1, \ldots, \beta_i, \ldots, \beta_{k'}\}$ digital output words are probed on well-selected target states. In principle only one initial driving state $|\psi(0)\rangle = |\psi_d\rangle$

is required, and depending on the way the output status of the gate is measured, there are 2 types of QHC logic gates. A first type of QHC gate is inspired by the qubit approach: a logic output is set up to "1" when the trajectory of the system reaches a given target state $|\phi_p\rangle$at $t = t_n$. This first approach is not very practical for the QHC because the population of the target state $\phi_p\rangle$ is in general a very fast time-dependent oscillation function which can reach a few THz or more of Ω_{dp} oscillation frequency.

In a second type of QHC gates, the Ω_{dp} Heisenberg–Rabi secular oscillation frequency between $|\psi_d\rangle$and $|\phi_p\rangle$ is used to encode the logical output. A very fast oscillation encodes for a logical "1" and a very slow oscillation for a logical "0" output. Since the $|\psi(t)\rangle = e^{-iHt/\hbar}|\psi(0)\rangle$ time evolution is the result of a linear superposition of many time-dependent components whose oscillation frequencies are simply the two-by-two differences of the $H(\alpha_1, \ldots, \alpha_i, \ldots, \alpha_k)$ eigenvalues, changing H via the $\{\alpha_1, \ldots, \alpha_i, \ldots, \alpha_k\}$ input words modifies the H eigenvalues and therefore the Ω_{dp} secular frequency. Then, positioning one H eigenvalue in resonance with the energy of the states $|\psi_d\rangle$and $|\phi_p\rangle$ will result in a pure Heisenberg–Rabi-like quantum oscillation, encoding for a logical output "1." However, if this eigenvalue is detuned from the exact resonance, then the Heisenberg–Rabi oscillation will be very slow corresponding to the logical "0" output. Let us here recall that when $|\psi_d\rangle$and $|\phi_p\rangle$ are coupled electronically to two metallic nanopads, the elastic current intensity I_{dp} passing from $|\psi_d\rangle$ to $|\phi_p\rangle$ through one of those manipulated H eigenvalue is simply proportional to the square of the corresponding Ω_{dp}. This offers a very practical way to measure a logical output "0" or "1." In QHC, a Boolean logic gate design finally results in the control of the position in energy of the H eigenvalues relative to some energy reference indexing a given output.

This chapter is organized as follow: In Sect. 2, we present the mathematical background of the QHC approach. We first show how the eigenvalues of a well-designed matrix can be shifted back and forth relative to zero and according to a given Boolean truth table when a small number of its matrix elements are modified. This eigenvalue repulsion effect is the first basic step of the QHC approach. Then, in Sect. 3, we will show how the eigenvalue repulsion effect can be used to construct all the universal symmetric two-input–one-output QHC logic gates. We also describe how to measure the logical output on a well-designed QHC quantum system. This is the second step of a QHC gate design. In Sect. 4, we demonstrate the atom-by-atom construction of the first prototypical atomic-scale QHC logic gate on a Si(100):H surface using UHV-STM vertical single atom manipulation. In Sect. 5, the electronic spectroscopy and the transport properties of the constructed QHC logic gate are analyzed using full valence scattering calculations, taking into account the Si(100):H surface, its supporting bulk, and the nanoelectrodes (the STM tip apex or the 2 planar metallic nanopads).

2 The Mathematics of a Quantum Hamiltonian Computing Boolean Logic Gate

In quantum mechanics, the eigenvalue repulsion effect is the result of the interaction of a minimum of two quantum states with the resulting property to increase the eigen energy difference between those two states. Considering for example the three-level quantum system presented in Fig. 1. The two states $|1\rangle$ and $|3\rangle$ of energy e are interacting with the state $|2\rangle$ of energy a via the two electronic couplings α and β and the state $|1\rangle$ is interacting with the state $|3\rangle$ via the electronic coupling k. In the corresponding $\{|1\rangle, |2\rangle, |3\rangle\}$ canonical basis set, the Hamiltonian of the system reads:

$$H_0(\alpha, \beta) = \begin{pmatrix} e & \alpha & k \\ \alpha & a & \beta \\ k & \beta & e \end{pmatrix} \tag{1}$$

Let us now assume that the two parameters α and β are the Boolean logical inputs and can respectively take only the value "0" or "1." Given a truth table having α and β as the two logical inputs, the Boolean logical output must be true when the Hamiltonian (1) has one eigenvalue equal to zero and false when (1) has no zero eigenvalue. This simple mathematical game can be solved by calculating the zero order term of the characteristic polynomial delivering the H_0 matrix eigenvalues. This comes from a well-know algebra theorem that the value of the zero order term of the secular equation governs the existence of an eigenvalue at zero. Therefore, the condition for at least one eigenvalue of the matrix (1) to be zero is simply that the determinant of $H_0(\alpha, \beta)$ equals to zero:

$$\Delta(\alpha, \beta) = ae^2 - e\left(\alpha^2 + \beta^2\right) - k^2 a + 2k\alpha\beta. \tag{2}$$

For a given set of structural parameters e, a, and k, the role of the (α, β) logical inputs is then to control the eigenvalues of matrix (1) relative to zero where zero was selected here as a reference energy indexing a given output and can be changed on purpose depending of the physical system used for implementing the QHC gate.

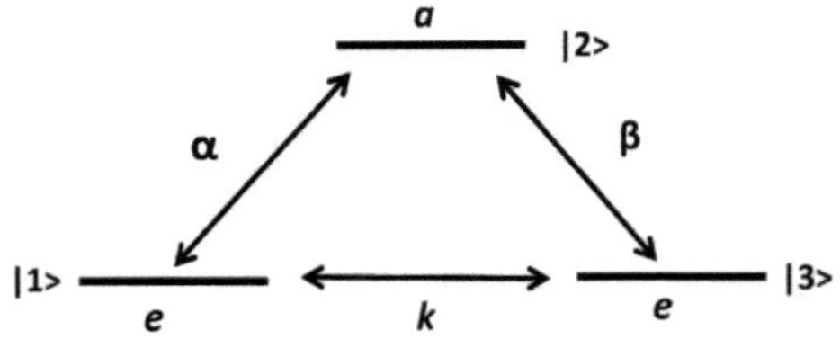

Fig. 1 Three-level quantum system used to construct all the universal two-input–one-output QHC logic gates. The (α, β) electronic couplings are the classical logical inputs of the QHC gate. The coupling k and the energies e and a are the structural parameters used to define the Boolean logic gate for a reading reference energy chosen here at $E = 0$

Table 1 Values of the k and a structural parameters and their corresponding zero order of the characteristic polynomial leading to the all six possible two-input–one-output symmetric QHC Boolean logic gates

Gate	Structural parameters	$\Delta\alpha\beta$
OR	$k = e/2,\ a = 4/(3e)$	$\Delta(\alpha,\beta) = e(1-\alpha^2-\beta^2+\alpha\beta)$
NOR	$k = 0,\ a = 0$	$\Delta(\alpha,\beta) = -e(\alpha^2+\beta^2)$
AND	$k = 0,\ a = e/2$	$\Delta(\alpha,\beta) = e(2-\alpha^2-\beta^2)$
NAND	$e = a = 0$	$\Delta(\alpha,\beta) = 2k\alpha\beta$
XOR	$k = 0,\ a = 1/e$	$\Delta(\alpha,\beta) = e(1-\alpha^2-\beta^2)$
NXOR	$k = e,\ a = 0$	$\Delta(\alpha,\beta) = e(2\alpha\beta-\alpha^2-\beta^2)$

This is the well-known eigenvalue repulsion effect pushing the (1) eigenvalues away from zero or not depending on the logical input values.

Starting from Eq. (2), one can now determine the e, a and k parameters to get all the OR, NOR, AND, NAND $H_0(\alpha, \beta)$ Boolean matrices and also the one for the composite XOR and the NXOR logic gates. Given a structural energy e, there are a few possible a and k families of values for each Boolean function, and we have selected in the following a homogeneous series for all them. The values for k and a structural parameters and their corresponding zero order term of the characteristic polynomial leading to the all six possible two-input–one-output symmetric Boolean logic gates are given in Table 1.

For example, for the composite XOR Boolean function we get: $k = 0$ and $a = 1/e$. In this case, the determinant of the $H_0(\alpha, \beta)$ matrix becomes:

$$\Delta(\alpha,\beta) = e(1-\alpha^2-\beta^2),$$

which is exactly equal to zero when the logical inputs α and β are exclusively equal to unity. The displacement in energy of the $H_0(\alpha, \beta)$ eigenvalues for the XOR Boolean logic function is shown in Fig. 2. $H_0(0, 0)$ [respectively $H_0(1, 1)$] has no zero eigenvalue. $H_0(1, 0)$ [respectively $H_0(0, 1)$] has one "0" eigenvalue.

Changing the α and β values is pushing one eigenvalue $H_0(\alpha, \beta)$ to reach exactly zero. This eigenvalue repulsion effect works for all the other OR, NOR, AND, NAND, and NXOR Boolean logic functions according to the structural parameters given in Table 1. Notice that an interesting property of those 3 × 3 Boolean matrices is also that the eigenvector associated with the zero eigenvalue (when it exists) has always a very large coordinate on the canonical vector $|2\rangle$. This will serve in the following section to construct a QHC Boolean logic gate with a well-defined output state.

The zero reference energy where the eigenvalue must be pushed to define a logical output "1" is one way to go from one gate to another by changing the values of the structural parameters as presented in Table 1. Another strategy is to fix the values of the structural parameters and then to change this reference energy to obtain all the six "two-input–one-output" gates. This second strategy of the QHC design offers a very practical way to construct QHC gates on the Si(100)H surface

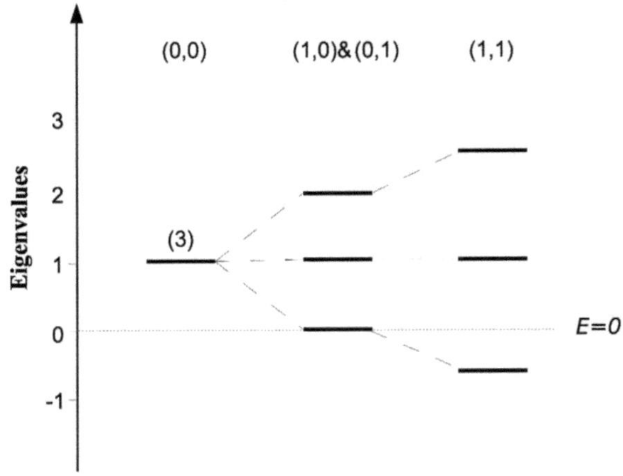

Fig. 2 Variation of the position of three eigenvalues of the $H_0(\alpha, \beta)$ 3 × 3 matrix as a function of the α and β logical inputs of the XOR Boolean function with $a = 1/e$, $k = 0$ and $e = 1$. The logical output is true when one of the $H_0(\alpha, \beta)$ eigenvalues is exactly equal to zero. Playing with the eigenvalue repulsion effect, this happens only for $(\alpha, \beta) = (0,1)$, and (1,0) leading to an XOR logic function

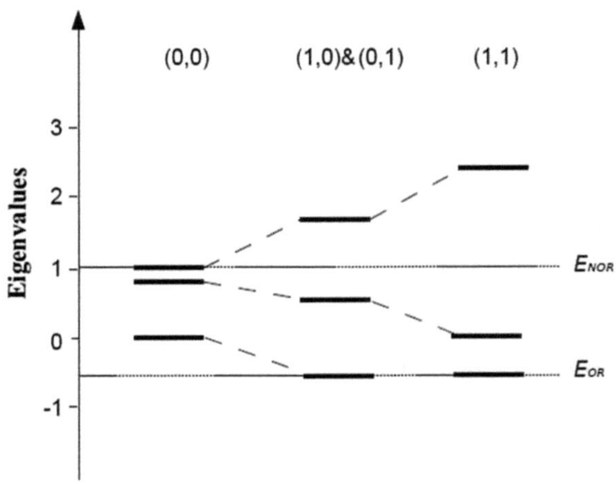

Fig. 3 Variation of the position of the three eigenvalues of the $H_0(\alpha, \beta)$ matrix as a function of the α and β logical inputs for $a = (e + \delta)/2$; $k = 4/(3e + \delta)$ and $e = 1 + \delta$ with $\delta = 1/2$. The NOR gate can be obtained by measuring the logical output at the reference energy $E = 1.0$ eV and the OR gate is obtained at $E = 0.5$ eV

since for a practical implantation, it is often the structural parameters which are determined (including the possible surface relaxations). As presented below, the reference energy to read the logical output status will be scanned in positive or negative voltages relative to the Fermi level of the supporting surface (see Sect. 4 for more details). One example is the QHC OR–NOR logic gates where after determining all the structural parameters using (1), its eigenvalues can be simply shifted in energy as presented in Fig. 3. As a consequence, the OR can be, for example, obtained at negative energy and the NOR at positive energy. This is exactly the QHC gate function presented in this chapter and constructed on Si(100)H.

3 The Universal Two-Inputs–One-Output QHC Logic Gate

To measure the logical output of a QHC gate with inputs (α, β), one simply has to construct a two-state system labeled here by $|m_1\rangle$, $|m_2\rangle$ of energy $E = 0$ which are only interacting with state $|2\rangle$ via the small coupling ε (see Fig. 4). This is the second step of the QHC gate construction. The QHC Hamiltonian is now a complete $H_5(\alpha, \beta)$ 5×5 matrix defined on the canonical basis set $\{|m_1\rangle, |m_2\rangle, |1\rangle, |2\rangle, |3\rangle\}$ by:

$$H_5(\alpha,\beta) = \begin{pmatrix} E & 0 & 0 & \varepsilon & 0 \\ 0 & E & 0 & \varepsilon & 0 \\ 0 & 0 & e & \alpha & k \\ \varepsilon & \varepsilon & \alpha & a & \beta \\ 0 & 0 & k & \beta & e \end{pmatrix}$$

where the reading head is the 2×2 top left matrix and the calculating block the 3×3 bottom right matrix. The reading matrix is simply here to measure if one of the eigenvalues of the 3×3 calculating matrix is resonating with its two eigenvalues E via the small coupling ε. Using this QHC 5×5 Boolean matrix, the complete dynamics of the system can be described as follow.

When this system is initially prepared at t = 0 in the non-stationary state $|m_1\rangle$ (i.e., $|\psi(0)\rangle = |m_1\rangle$), it will spontaneously evolve in time toward the target state $|m_2\rangle$ and though the state $|2\rangle$ according to the Schrödinger equation:

$$i\hbar \frac{\mathrm{d}}{\mathrm{d}t}|\psi(t)\rangle = H_5(\alpha,\beta)|\psi(t)\rangle.$$

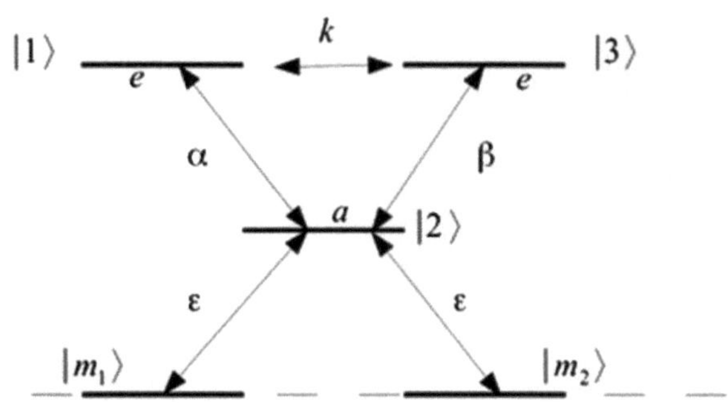

Fig. 4 Quantum graph of the $H_5(\alpha, \beta)$ QHC represented here on its $\{|m_1\rangle, |m_2\rangle, |1\rangle, |2\rangle, |3\rangle\}$ canonical basis set with the calculating block defined on $|1\rangle$, $|2\rangle$, and $|3\rangle$ and the reading head defined on $|m_1\rangle$ and $|m_2\rangle$. The (α, β) electronic couplings are the classical logical inputs of the QHC gate. The coupling k and the energies e and a are the structural parameters used to define the Boolean logic gate for a reading reference energy chosen here at $E = 0$, i.e., the energies of $|m_1\rangle$ and $|m_2\rangle$

Due to the small coupling ε, one oscillation frequency labeled here Ω (α, β) mainly governs the time evolution from the initial state $|m_1\rangle$ to the target state $|m_2\rangle$. In particular, it is easy to show that Ω (α, β) will be very fast when one eigenvalue of the 3 × 3 calculating block is in exact resonance with the energy of the states $|m_1\rangle$ and $|m_1\rangle$, namely equal to zero. However, if this eigenvalue is detuned in energy relative to zero, this oscillation will be very slow. This exactly corresponds to the mathematical game described above: The output of the preceding Boolean function is true only when one eigenvalue of the 3 × 3 calculating block is exactly equal to zero. One example is the XOR logic gate whose complete dynamics is shown in Fig. 5. To certify the correct functioning of this logic gate, we have numerically solved the time-dependent Schrödinger equation by preparing the initial state $|\psi(0)\rangle = |m_1\rangle$ and then calculating the population of the target state $|m_2\rangle$: $P(t) = |\langle m_2|\psi(t)\rangle|^2$.

As presented in Fig. 5, the population of the target state is rapidly oscillating only when the logical inputs α and β are exclusively equal to "1" leading to the

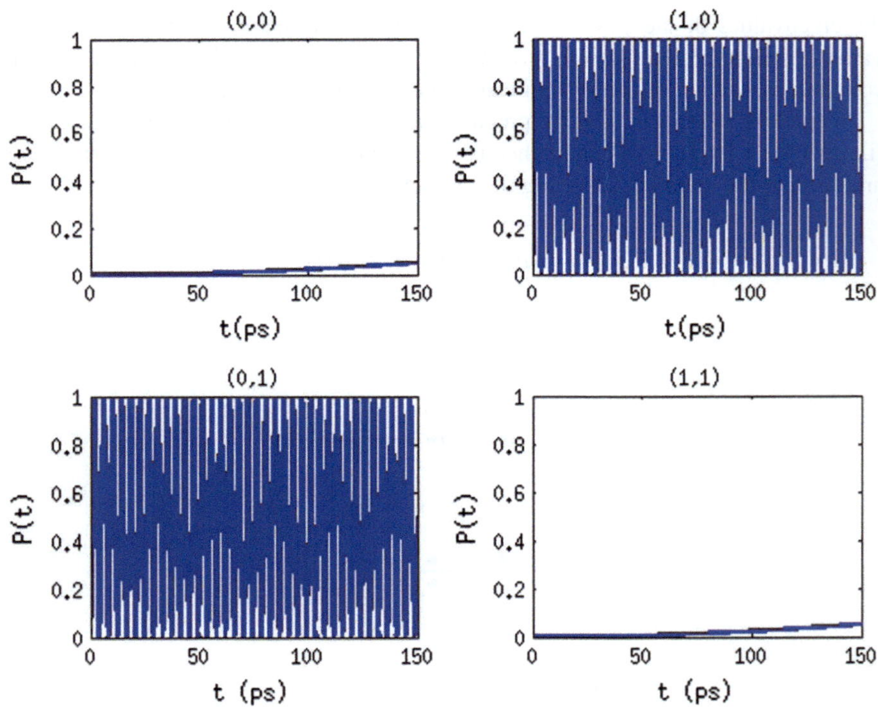

Fig. 5 Exact time-dependent variations of the population $P(t) = |\langle m_2|\psi(t)\rangle|^2$ as a function of the logical input configuration (α, β) (indicated on each curve) obtained by numerically solving the time-dependent Schrödinger equation for the specific choice of $e = a = 1$ eV, $k = 0$ eV and $\varepsilon = 0.001$ eV, and after preparing the quantum system in the non-stationary initial state $|m_1\rangle$. The Heisenberg–Rabi oscillation is fast only when the logical inputs are exclusively equal to "1" corresponding to the XOR logic function

XOR logic function. Notice that the exact meaning of "fast" and "slow" depends on the nature of the quantum system whose typical oscillation frequencies can range from Gigahertz to Terahertz. For example, for the case of the XOR function and for the specific choice of $e = a = 1$ eV and $\varepsilon = 0.001$ eV, the logical output "1" is obtained when the secular Heisenberg–Rabi oscillation Ω (α, β) is maximum, corresponding to a frequency of 0.23 THz. The logical "0" is encoded for a minimum value of $\Omega = 0.48$ GHz.

4 Construction of Surface QHC Atomic Logic Gates

One way to construct the QHC atomic logic gates described above is by extracting single H atoms from a Si(100):H surface using a scanning tunneling microscope (STM) tip in a vertical atom manipulation operation mode. In this case, each created dangling bond (DB) can formally be mapped with one of the canonical basis set vectors used to define a QHC Boolean matrix. But due to the valence electronic structure of each Si surface atom, this is only partially correct since the canonical basis set in which the QHC gate will be constructed on a Si(100):H surface contains at least three quantum states per site instead of one for the QHC quantum graph used in Fig. 4. Different Boolean logic operations may be defined depending on the well-determined surface network of the created DBs. In order to design complex QHC logic gates on the Si(100):H surface, it was first necessary to analyze how the quantum state of the DB can be manipulated by neighboring DBs that act as logic inputs, following the QHC approach. Therefore, we decided to construct a very simple gate with a minimum number of DBs to analyze in detail with scanning tunneling spectroscopy (STS) how and whether a simple QHC gate works. An n-type phosphorus doped Si material was used as the substrate. The top Si(100):H surface was prepared in situ by conventional annealing followed by flashing as presented in Chapter "Surface Hydrogenation of the Si(100)-2×1 and Electronic Properties of Silicon Dangling Bonds on the Si(100):H Surfaces". The experiments were performed on our Omicron LT-UHV-STM at 4 K.

Single hydrogens were removed using an STM tip: one for the central active DB site and one for each logical input (see Fig. 6 for the exact position of the logical input sites). The atomic structures of the reference Si(100):H surface (no DBs) and the three input configurations are presented in Fig. 6 together with the corresponding empty-state LT-UHV-STM images.

The design of this surface QHC DB logic gate exactly follows the QHC graph in Fig. 1 with the central quantum state $|2\rangle$ with energy "a" being the central DB in the implementation in Fig. 6. For the inputs, the following convention was used: The removal of one H atom which creates a single DB encodes a logical "0," while the addition of one H atom to the single DB encodes a logical "1" input. This defines the input states $|1\rangle$ and $|3\rangle$ in Fig. 1 via their through space (p_z orbitals) or

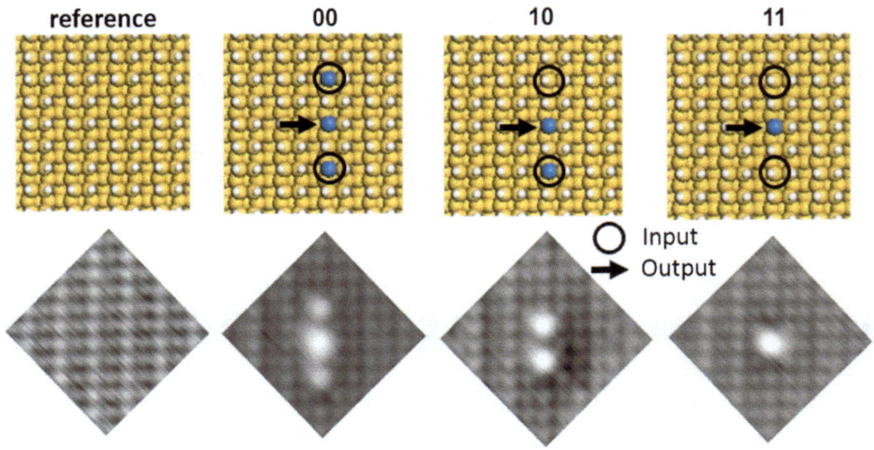

Fig. 6 Surface atomic structure of the reference Si(100):H surface and the three logical configurations of the NOR/OR QHC logic gate. The input and output sites are indicated with circles and arrows, respectively. The corresponding empty-state LT-UHV-STM images are presented below each surface structure ($V = +1.5$ V, $I = 20$ pA)

through surface (p_x, p_y orbital) electronic coupling. The inverse convention can also be chosen. The present one is corresponding to the molecule latch inputs which will be discussed in Chapter "Molecule-Latches in Atomic Scale Surface Logic Gates Constructed on Si(100)H". As described later, these DB configurations demonstrate NOR or OR QHC logic behavior, depending on the bias voltage used for the electronic spectral output measurement.

The electronic dI/dV characterization of the QHC logic gates shown in Fig. 6 was performed using a vertical STS dI/dV technique using the lock-in instrument of the LT-UHV-STM system (Fig. 7). In absence of an available atomic-scale planar interconnection technology, the output status of any atomic-scale surface logic gate has to be determined by pointing the STM tip at the preselected output DB (central DB in this case) and by measuring the direct tunneling current intensity through this DB. According to the QHC design, the quantum information propagating along the DB atom circuit will reach the output atom depending on the status of the inputs. The gate truth table can therefore be tested by dI/dV spectroscopy, characterizing how the surface states of this quantum circuit are shifted according to the binary input status.

At the surface conduction band edge (+1.3 V), a tunneling resonance peak is observed for the "00" input, whereas no peak is observed for "10" and "11" input configurations, demonstrating a NOR gate behavior (Fig. 7). On the other hand, at the valence band edge (−1.8 V), a large resonance peak is observed for "10" and "11" input configurations, but the peak is much smaller for "00" input configuration, demonstrating an OR gate behavior (Fig. 7).

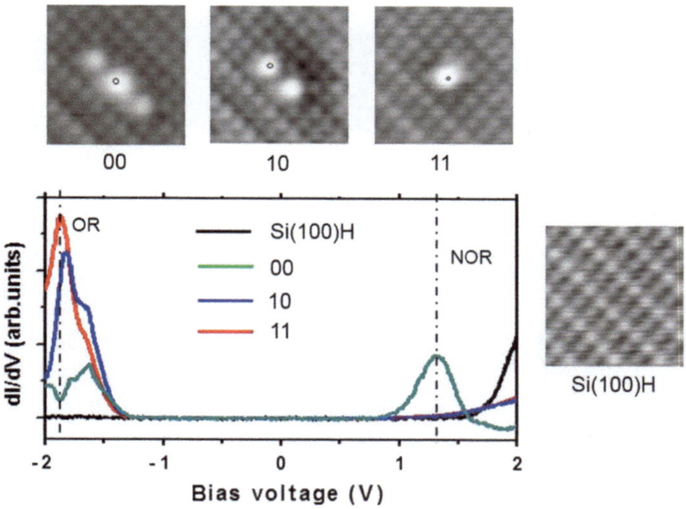

Fig. 7 STM images and dI/dV spectra of the reference Si(100):H surface and three logic configurations, demonstrating a NOR gate behavior at positive bias voltage and an OR gate behavior at negative bias voltage in a spectral mode of output status measurement

The geometry and the electronic properties of these DB configurations will be interpreted in detail in Sect. 5, where we compare the experimental measurements with calculated results. Nonetheless, these experimental results show that the QHC DB surface logic gates can be constructed on a Si(100):H surface using a STM tip, and the STS dI/dV technique allows for their electronic characterizations and determination of their logic gate behavior.

To check the influence of the through-bond coupling between the input (edge) DB and the calculating central DB, another geometry has been investigated, where we extracted the H atom on an adjacent dimer row instead of the same dimer row (Fig. 8). The STM image for each configuration is shown in Fig. 8.

The I–V curves for the three configurations are very similar in this case (Fig. 9), showing that this lateral shift in the logical input location was enough to largely reduce the electronic interactions between the DBs. The central (output) DB electronic state is not affected by the presence of the lateral input DBs, and no logic gate response is observed. This confirms the QHC gate design in Fig. 6 and the possibility to fine-tune the through-surface interactions between DB by a proper inter-DB surface distance.

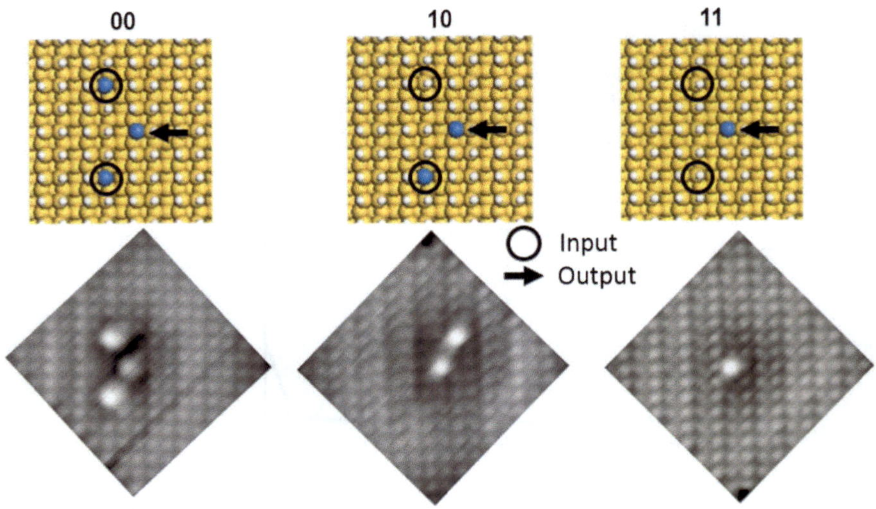

Fig. 8 Surface atomic structures and corresponding empty-state LT-UHV-STM images ($V = +1.5$ V, $I = 20$ pA) of the three configurations with the logical input location at the adjacent dimer row. The input and output sites are indicated with circles and arrows, respectively. No logic gate behavior was observed for these DB configurations (see Fig. 9)

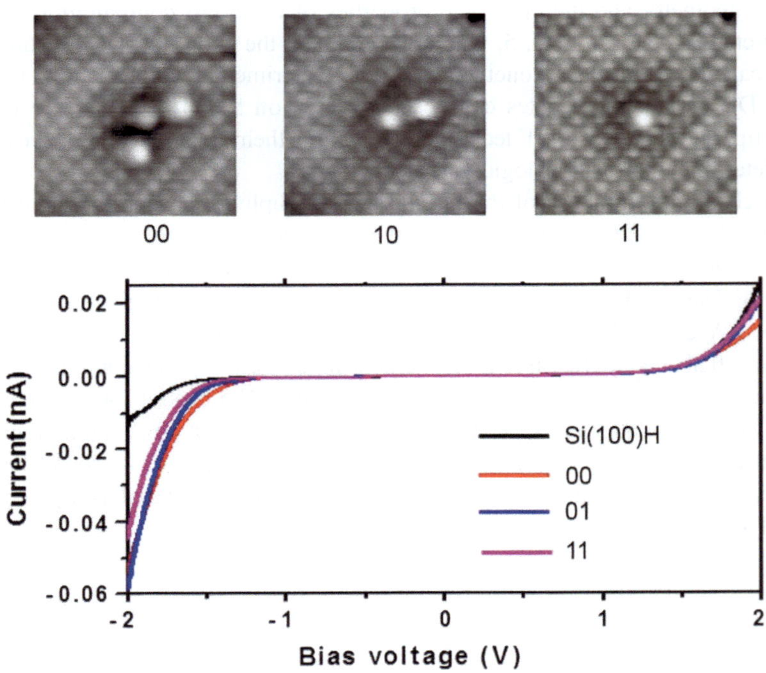

Fig. 9 STM images and *I*–*V* spectra of the three configurations shown in Fig. 8. *I*–*V* curves for all three configurations are very similar, and no logic gate behavior is observed for these configurations

5 Theoretical Calculation of the Electronic Properties of a Simple DB Logic Gate on a Si(100):H Surface

The electronic properties of the simple QHC DB logic gate constructed on a Si(100):H surface as presented in the previous section were investigated using density functional theory (DFT) and quantum transport simulations. These calculations evaluate the relationship between the simple QHC graph in Fig. 1 and the complete electronic structure of the implementation on a Si(100):H surface using DB states as shown in Fig. 6. The Si(100):H surface was modeled as a five-layer slab with the top and bottom surfaces passivated with hydrogen. 1, 2, and 3 DB structures were created by removing the corresponding number of *H* atoms from the surface. Electrons were added to the unit cell to model the doping in the *n*-type Si. The DBs were assumed to be charge saturated (i.e., one electron was added per DB), and the surface geometries were optimized with DFT with the Perdew–Burke–Ernzerhof (PBE) functional [1], as implemented in the Vienna ab initio simulation package (VASP) [2–5].

The surface Green function matching (SGFM) method with the extended Hückel molecular orbital (EHMO) Hamiltonian was used to calculate the STM images and the *T(E)* transmission spectra [6]. The structure of our STM model is shown in Fig. 10a. The tip-induced band bending (TIBB) was taken into account by shifting the energies

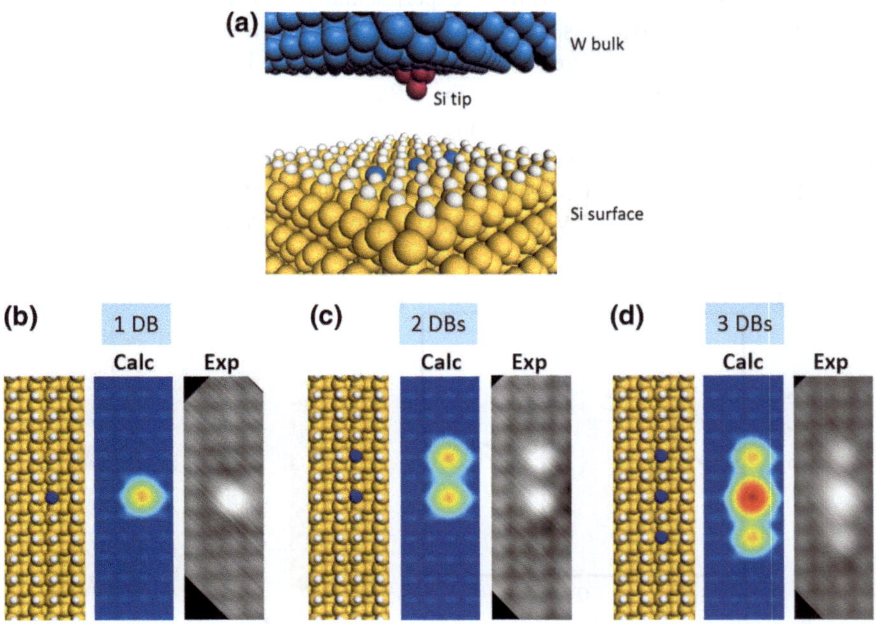

Fig. 10 **a** STM structure used in the image and *T(E)* calculations. **b–d** Atomic structure of 1, 2, and 3 DB configurations with corresponding calculated and experimental empty-state STM images. (Calculated: $I = 20$ pA, $V = +0.5$ V; Experiment: $I = 20$ pA, $V = +1.5$ V. The difference in bias voltage is due to the difference in tip-induced band bending (TIBB))

of the surface states with respect to the bulk bands (see Chapter "Band Engineering of the Si(001):H Surface by Doping with P and B Atoms" for more details).

First, the structures of the surfaces with DBs were confirmed by comparing the calculated empty-state STM images with the experimental STM images (Figs. 10b–d). For 1 and 2 DBs, the DB appears round and symmetric. The DB is slightly buckled up because the addition of an extra electron per DB due to *n*-type doping causes the DB Si atom to be more sp_3 hybridized. On the other hand, for 3 DB case, the central DB appears larger and brighter compared to the DB at the edge, implying that the *T(E)* is larger when the STM tip is above the central DB than above the edge DBs. This is because the buckling angle of the central DB is slightly smaller than that of edge DBs due to the reduction in the charge on the inner DB caused by Coulomb repulsion (see Chapter "Band Engineering of the Si(001):H Surface by Doping with P and B Atoms" for more details), and the energy positions of the central and edge DB states are different, as discussed more later.

Note that the bias voltages used in the calculation and in the experiment are different because of TIBB. Since it is challenging to determine the exact amount of TIBB in the experiment and our goal is to analyze the trend in the *T(E)* spectra and not to reproduce exactly the experimentally measured STS spectra, it is sufficient to use the approximated TIBB based on the observed STM images and STS spectra (upward bending of about +0.2 V was used in this case). Nevertheless, the STM images calculated at small positive bias (corresponding to the conduction band edge) are in good agreement with the experimentally measured STM images (also at conduction band edge), as shown in Figs. 10b–d.

Next, the *T(E)* spectra were calculated for the 1, 2, and 3 DB configurations with the STM tip positioned above one of the DBs (Fig. 11) as used experimentally in the d*I*/d*V* measurements presented in Fig. 7. For the 1 and 2 DB cases, a large *T(E)* peak is observed in the negative bias range, which corresponds to the DB state. The DB state appears in the negative bias range due to the downward TIBB (an approximate value of −0.7 V was used) for the *T(E)* calculation for the valence

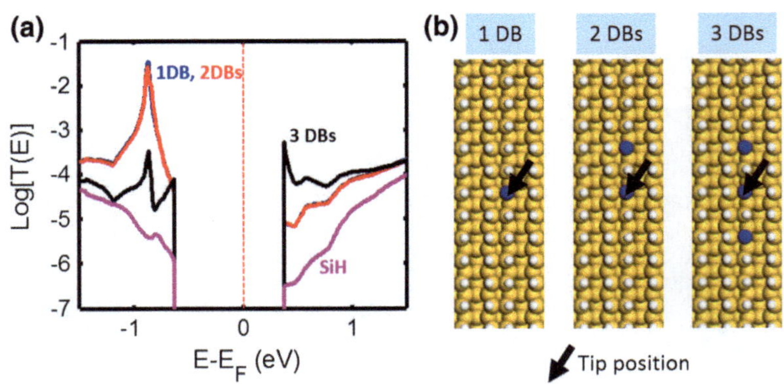

Fig. 11 **a** *T(E)* spectra of Si(100):H surface with 1, 2, 3 DBs and with no DB (SiH). **b** Atomic structures of Si(100):H surface with 1, 2, and 3 DBs. The tip position for each case is indicated by an arrow

band range. The splitting of the *T(E)* peak in the 2 DB case is very small, and the position of the *T(E)* peaks is very close for 1 and 2 DB cases. This shows that the buckled structure and the energy position of the 2 DBs are the same, and the coupling between the two DBs in 2 DB case is weak.

There is no apparent *T(E)* peak in the positive bias range for the 1 and 2 DB case. The *T(E)* spectrum for the 3 DB case, however, is significantly different from that of the 1 and 2 DB cases. In the negative bias region, there is a peak at the same energy with the DB state in 1 and 2 DB cases, but the peak is much smaller in 3 DB case. The small *T(E)* peak likely corresponds to the state of the edge DBs. The peak is smaller because the STM tip is positioned above the central DB and not above the edge DB. If the STM tip is placed above the edge DB, this peak becomes larger (not shown), confirming that this peak is indeed associated with the edge DB state. This also agrees with the similar buckling angle for the edge DBs in 3 DB case (+7.3°) and for the 1 and 2 DB cases (+7.3°). In the positive bias region, there is a small peak at the conduction band edge, which is not observed in 1 and 2 DB cases, suggesting that the *T(E)* peak at conduction band edge corresponds to the DB state of the central DB in 3 DB case (above which the STM tip is positioned). As mentioned before, the energy position of the central DB state is different from that of the edge DBs because of its smaller buckling angle (+5.6°). Specifically, the central DB state is higher in energy than the edge DB states. Note that the energy difference between the edge DB states and the central DB state is not necessarily larger than the Si band gap. In this case, the upward TIBB of about +0.2 V was used for *T(E)* calculation for conduction band range, which results in the appearance of the central DB state at the conduction band edge. However, the exact position of the peak may vary depending on the amount of TIBB. Note that in this case, the tail of the central DB state also appears in the valence band edge due to the downward TIBB for negative bias region, but depending on the amount of TIBB, this peak may or may not be seen in the valence band edge. The trend observed in these *T(E)* spectra generally agrees with the experimentally measured dI/dV spectra shown in Fig. 8, demonstrating the OR and NOR gates at negative and positive bias, respectively, also for the realistic electronic structure. The 3 DB structure with the input (edge) DBs on the adjacent dimer row (Figs. 8 and 9) was not analyzed in the calculation, but the STM images and the IV curves in Fig. 9 suggest that in this case, the three single DBs are isolated from each other and have similar buckling angles. Since the energy positions of all DBs are expected to be the same, no logic behavior can be observed.

In order to explore the possibility of using such a simple QHC logic gate structure in a planar configuration using metallic nanopads, the *T(E)* spectra for the same DB QHC circuit and configurations (five-layer slab) were calculated for a planar configuration (Fig. 12) and compared with those calculated in the STM configuration as presented in Fig. 11. In the planar configuration, the DBs were contacted by Au pads placed on the Si(100):H on the surface instead of by STM tip (Fig. 12c), and the *T(E)* is calculated between the Au pads. The DB states are observed within the Si(100):H surface band gap, and the *T(E)* in the gap, where there are no *T(E)* peaks, depends on the leakage current through the Si(100):H

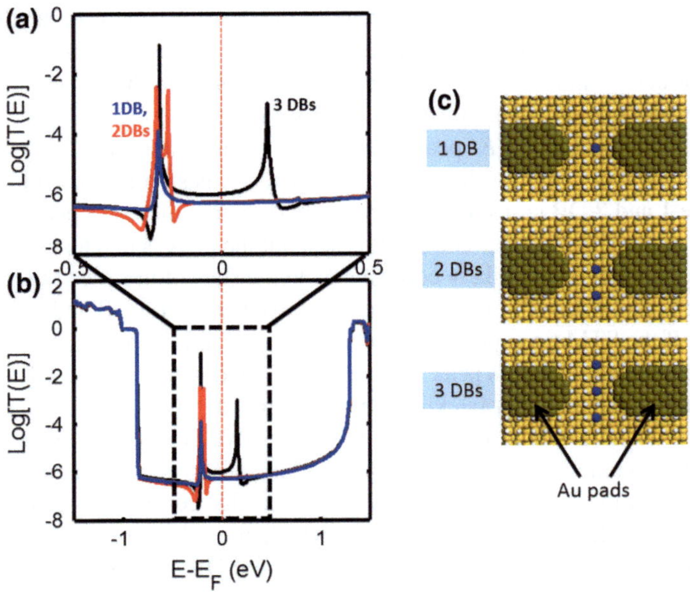

Fig. 12 **a** *T(E)* spectra for the Si(100):H surface with 1, 2, 3 DBs and without DBs (Si(100):H). **b** Atomic structures of the Si(100):H surface with 1, 2, and 3 DBs. The Au pads are indicated by arrows

surface. Figure 12b shows the *T(E)* in the large energy range to show the edges of the band gap, and Fig. 12a shows the zoom in for the energy range where the *T(E)* peaks are observed. Similar to the STS case, the *T(E)* peaks corresponding to the DB states are observed below the Fermi energy for 1 and 2 DB cases, whereas the peaks are observed both below and above the Fermi energy for 3 DB case, corresponding to the edge and central DB states, respectively. Note that in this case, no band bending is assumed, so the energy difference between the edge and central DB states is much smaller than that observed in the *T(E)* calculation for STM configuration. The actual energy difference needs to be confirmed by experimental measurements in the same planar configurations, but this is beyond the scope of this work.

Although the exact energy difference is unknown, the edge DB states are likely to appear below the Fermi energy and the central DB states are likely to appear above the Fermi energy, demonstrating a NOR gate at positive bias (around +0.16 V in this case). For the negative bias range, however, the intensity of the *T(E)* peaks is similar for the 1, 2, and 3 DB cases, and no OR gate behavior can be observed. This is because unlike in the STM configuration, where the STM tip is positioned above the central DB which carries, according to the QHC graph in Fig. 1, the output status of the gate, in the planar structure all the DBs are contacted by the Au pads. The Au pads are not only electronically coupled to the central DB state, but also to the edge DB states. In order to make an OR gate similar to the

STM configuration, the coupling between the Au pads and the edge DBs needs to be reduced by, for example, making the Au nanopad very sharp and focused on the central DB.

6 Conclusions

Starting from the formal quantum graph topology of a simple two-input–one-output QHC Boolean logic gate, a dangling bond QHC gate was atom-by-atom constructed on a Si(100):H surface using the LT-UHV-STM microscope in a vertical atom manipulation mode. Each state of the formal QHC quantum graph was attributed to a single DB, knowing however that this one-to-one mapping between DB quantum states and the QHC quantum state canonical basis set is not exact. According to the QHC graph topology, the Si(100)H implementation can function as a NOR or a OR Boolean logic gate depending on the reference energy used to read the corresponding logical output. The experimental measurement on the central DB atom shows NOR gate behavior at a positive bias voltage of 1.3 V and OR gate behavior at a −1.8 V bias voltage. The QHC function was confirmed in another experiment by slightly decoupling the 2 DB input states from the central output DB. Full electronic dI/dV calculations reproduce the LT-UHV-STM dI/dV STS spectra, confirming the NOR and OR QHC gate function depending on the bias voltage. The simulation of a fully planar configuration of this QHC gate indicates that the two planar Au nanoelectrodes are too close to capture only the electronic status of the central output DB as required in the formal QHC design.

Interestingly, while the QHC graph topology provides a good starting point to guide the construction of a real QHC logic gate, it turns out that the detailed functioning of the OR–NOR gate on the theoretical canonical basis set does not correspond to the functioning of the gate in its full valence electronic structure including the supporting surface electronic structure and all the through-surface electronic coupling. Therefore, this simple 3 DB surface Si(100):H construction is a very instructive step on the way toward an atomic-scale QHC logic gate, but for a realistic planar implementation, a more complex QHC quantum graph must be developed taking into account the actual through-surface and through-space electronic coupling between the DB states. Such couplings are currently not included in the QHC theoretical canonical (and orthogonal for simplicity) basis set.

Acknowledgements We acknowledge the Agency of Science, Technology, and Research (A*STAR) for funding provided through the Visiting Investigatorship Programme Atom Technology project 1021100972, and through the AtMol integrated project contract number 270028 from the European Commission. We also acknowledge the A*STAR Computational Resource Centre (A*CRC) for computational resources and support.

References

1. Perdew, J.P., Burke, K., Ernzerhof, M.: Generalized gradient approximation made simple. Phys. Rev. Lett. **77**, 3865 (1996). doi:10.1103/PhysRevLett.77.3865
2. Kresse, G., Hafner, J.: Ab initio molecular dynamics for liquid metals. Phys. Rev. B **47**, 558 (1993). doi:10.1103/PhysRevB.47.558
3. Kresse, G., Hafner, J.: Ab initio molecular-dynamics simulation of the liquid-metal-amorphous-semiconductor transition in germanium. Phys. Rev. B **49**, 14251 (1994). doi:10.1103/PhysRevB.49.14251
4. Kresse, G., Furthmüller, J.: Efficient iterative schemes for ab initio total-energy calculations using a plane-wave basis set. Phys. Rev. B **54**, 11169 (1996). doi:10.1103/PhysRevB.54.11169
5. Kresse, G., Furthmüller, J.: Efficiency of ab-initio total energy calculations for metals and semiconductors using a plane-wave basis set. Comput. Mater. Sci. **6**, 15 (1996). doi:10.1016/0927-0256(96)00008-0
6. Cerdá, J., Van Hove, M.A., Sautet, P., Salmeron, M.: Efficient method for the simulation of STM images. I. Generalized green-function formalism. Phys. Rev. B **56**, 15885 (1997). doi:10.1103/PhysRevB.56.15885

The Design of a Surface Atomic Scale Logic Gate with Molecular Latch Inputs

Francisco Ample, Omid Faizy, Hiroyo Kawai and Christian Joachim

Abstract Two-input/one-output atomic scale Boolean logic gates are presented which are supposed to be atom by atom STM constructed on an Si(100) H-passivated surface. For those relatively simple logic gates, semi-classical and quantum Hamiltonian computing (QHC) architectures are compared showing how the QHC design avoids long atomic scale wiring compulsory for any classical or semi-classical "à la Shannon" electric circuit design. For a more realistic approach, the simple double hydrogen switching elements used for the logical inputs were substituted by mechanical molecule latch inputs. This does not transform the logical Boolean truth tables of the surface-implanted logic gates. For both the semi-classical and the QHC designs, the four standard known Boolean logic gates: AND, NOR, NAND and OR can be formally designed with an exception of the OR for the QHC approach. On the surface, AND, NOR, NAND and OR can be implanted in a semi-classical way and NOR and OR in a QHC way. For QHC, the re-emerging of the OR is due to through-surface electronic couplings, which is inducing a dependence between the inputs and the structural parameters of the QHC gates.

F. Ample · H. Kawai
Institute of Materials Research and Engineering, 2 Fusionopolis Way, Innovis, #08-03, Singapore 138634, Singapore

O. Faizy · C. Joachim (✉)
GNS & MANA Satellite, CEMES-CNRS, 29 rue J. Marvig, 31055 Toulouse Cedex, France
e-mail: joachim@cemes.fr

C. Joachim
International Center for Materials Nanoarchitectonics (WPI-MANA), National Institute for Materials Science (NIMS), 1-1 Namiki, Tsukuba, Ibaraki 305-0044, Japan

M. Kolmer and C. Joachim (eds.), *On-Surface Atomic Wires and Logic Gates*, Advances in Atom and Single Molecule Machines,
DOI 10.1007/978-3-319-51847-3_9

1 Introduction

Embedded in a single conjugated molecule or constructed at the surface of a semiconductor wafer, there are two ways to design an atomic scale logic gate: semi-classical or quantum. In a semi-classical approach, both the molecule and the surface atomic scale circuit look like a standard electronic circuit [1]. In a quantum Hamiltonian computing (QHC) approach, the quantum system (the molecule or a surface arrangement of dangling bonds) defines specific quantum states which can be manipulated by the inputs with no ressemblance with an electronic circuit ([1] and see Chapter "Quantum Hamiltonian Computing (QHC) Logic Gates").

For intramolecular semi-classical electronic circuits, the general intramolecular electronic circuit rules for tunneling and ballistic regimes of transport have already been established [2]. Semi-classical circuit can also be constructed by extracting surface atom one after the other at the surface of a semiconductor. The electronic conductance from contact pad to contact pad of those atomic scale surface circuits can be calculated using the semiempirical N-ESQC technique to determine the best design. Here, N-ESQC stands for N electrodes elastic scattering quantum chemistry calculations [3].

In this chapter and at the surface of a passivated semiconductor, the design of "monolithic" Boolean logic gates resulting from the construction of a surface atomic scale circuit is presented following either a semi-classical or a full quantum approaches. For the semi-classical approach, we have simply supposed that the topology of the logic circuit is the same at the atomic scale as compared to macroscopic logic electronic circuits. In this case, the circuit is constructed atom by atom with meshes, nodes and by introducing switches when necessary along the circuit to perform the Boolean logic function. For a quantum approach, we have privileged the QHC design already working well inside a single molecule [1, 4], i.e., we will show in this chapter how to play with the repulsion quantum level effect and with destructive quantum interferences within an atomic scale circuit on an Si(100)H surface with a more complex atomic scale configuration than the simple QHC gate presented in Chapter "Quantum Hamiltonian Computing (QHC) Logic Gates" above.

For QHC and on the Si(100)H surface, the logical input strategy will formally follow the NOR starphene molecule experimental design where one single atom was manipulated with the tip of a scanning tunneling microscope (STM) and this per classical input [3]. On Si(100)H, atomic scale logical inputs are more complex to implement as compared to single atom logical input on a molecule because the number of surface states involved is generally much larger. For example, a simple Si dangling bond created by STM tip vertical manipulation on an Si(100)H surface must be represented by a 4×4 matrix and not by a single electronic coupling as in the molecular QHC design. A systematic brute force optimization will be first used in this chapter to determine the location of the required "calculating" dangling bond on an Si(100)H surface for the measured conductance of the gate to follow a Boolean logic truth table.

As presented in this chapter, both the classical circuit and QHC approaches are leading to the design of two-input/one-output AND, NOR, OR and NAND Boolean logic gates on a Si(100)H surface. In this chapter and for both approaches, we have also started to introduce a more realistic classical mechanical input design by the optimization of a molecular latch logical input per digital input using an acetophenone molecule. Those molecular latches will be further explored in Chapter "Molecule-Latches in Atomic Scale Surface Logic Gates Constructed on Si(100)H". Those molecular latches can be considered as an effort to miniaturize the input keyboard as close as possible in scale to the calculating unit, a new practice that is very difficult to implement with the now well-known qubit approach of quantum computers.

2 A Semi-classical Surface Dangling Bond Logic Design on Si(100)H

All the OR, NOR, AND, NAND two-input and one-output Boolean logic gates can be designed following the new circuit design rules demonstrated in [2] and verified for the first time experimentally for the mesh circuit law in 2012 inside a single molecule [5]. They were designed on the Si(100)-(2×1)-H surface. Their interconnections to the macroscopic world are simulated by metallic nano-pads supposed to be physisorbed also on this Si(100)H surface [6]. To simplify the design, the logic inputs are given by H saturating and un-saturating two surface Si dangling bonds (DBs) at a time. More realistic input latching using a single molecule for inputting the logic data on those circuits is presented for one example at the end of this section and more generally in Chapter "Molecule-Latches in Atomic Scale Surface Logic Gates Constructed on Si(100)H". They are all supposed to be driven by electronic tunneling inelastic effects. Quantum circuit design rules together with semiempirical N-ESQC transport calculations were used to determine the output current intensity of the proposed logic gates when also interconnected to the metallic nano-pads using surface atomic scale wires. Those rules are based on a semi-classical approach, decomposing the central circuit Hamiltonian in parts representing the interconnection wires or the input switches of the logic gates.

The building blocks of our atomic scale Boolean logic gates are elementary atomic scale switches composed of 3 DB dimers in a row contacted by Au nano-pad at each end. The logic inputs "1" and "0" correspond to the action of hydrogenating and dehydrogenating the central DB dimer, respectively. For the short atomic wire Fig. 1a, addition of two H atoms on the central dimer corresponds to a logic input "1" (Fig. 1b), resulting in an inverter. Alternatively, one can also start from the junction shown in Fig. 1c and add two H atoms to the Si DB dimer adjacent to the central dimer to achieve a logic input "1" (Fig. 1d), which results in a follower. As shown in the truth table in Fig. 1, for the inverter, a logic input "1" turns the device "OFF," whereas for the follower, logic input "1" turns the device "ON."

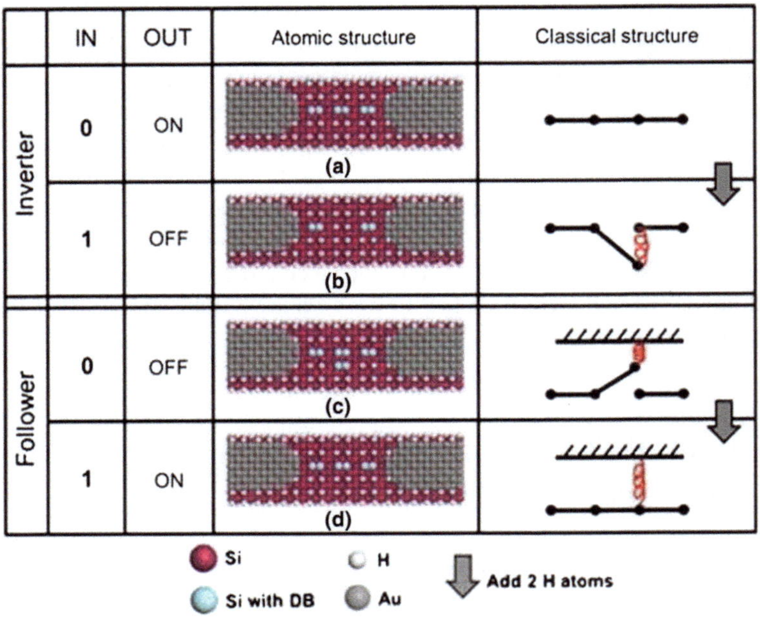

Fig. 1 Atomic structures for the two atom switches designed on a Si(100)-(2×1)-H surface. In going from structure **a** to **b**, the addition of two H atoms to the central dimer encodes a logic "1" input, leading to an inverter. In going from structure **c** to **d**, the addition of to the lateral dimer also encodes a logic "1" input, leading to a follower. The classical equivalent electromechanical structure for each structure is also shown

In this semi-classical approach, the generalization of the Kirchhoff's law of electrical circuits for quantum circuits was used to design semi-classical dangling bond logic gates [7]. For example and in the absence of decoherence, the interconnection of two quantum electronic pathways in series with elementary transmission coefficients $T_X(E)$, $T_Y(E)$ leads to the overall $T_{\text{series}}(E)$:

$$T_{\text{series}}(E) = T_X(E) \cdot T_Y(E) \tag{1}$$

On the other hand, when the atomic devices are connected in parallel, and when the nodes of the circuit are included inside the quantum system, i.e., again with no decoherence, the overall $T_{\text{parallel}}(E)$ is given by (in the tuneling regime [8] if there is also no phase shift between each branch of the circuit):

$$T_{\text{parallel}}(E) = T_X(E) + T_Y(E) + 2\sqrt{T_X(E) \cdot T_Y(E)} \tag{2}$$

Based on these rules, the truth table for each AND, NOR, NAND and OR logic gates and their corresponding atomic scale structures can be obtained (Fig. 2).

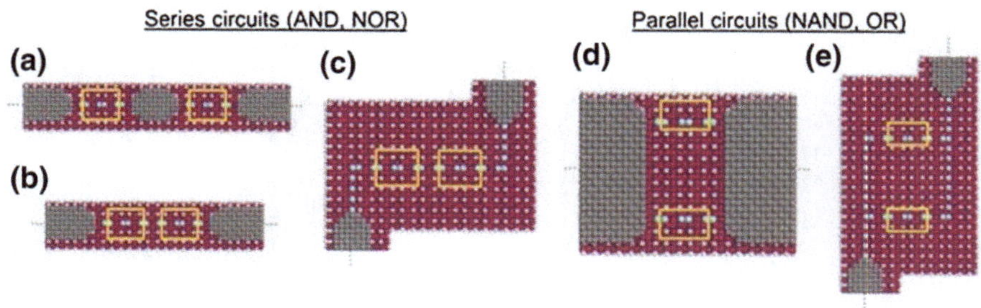

Fig. 2 Different ways of connecting the two surface atomic switches in series (**a–c**) and in parallel (**d**, **e**) on a Si(100)-(2×1)-H surface

Those four Boolean logic gates can also be constructed by interconnecting the atomic scale inverter and follower devices in series or in parallel following a C. Shannon like approach but using the new quantum circuit rules (1) and (2). For the circuits in series, the two switches can be connected via a central nano-island (Fig. 2a) or connected directly (Fig. 2b). But the gate can also be contacted by two DB wires parallel to the dimer rows [9] and then by Au nano-pads (Fig. 2c). For the circuits in parallel, the two switches can be connected by one large Au nano-pad on each side (Fig. 2a), or can be connected using surface DB atomic wires and then to Au nano-pads (Fig. 2b). On all those circuits, the locations for the two inputs in Fig. 2 are indicated using transparent yellow squares.

The electronic $T(E)$ transmission spectrum for each circuit and input logical configuration was calculated using $N = 2$ ESQC. Then, the I–V curves were obtained for each gate using the generalized Landauer formula (see Ref. 2) supposing always that one of the Au nano-pads which is connected to the ground to beneficiate from the complete variations of the $T(E)$ spectra, the voltage drop occurring in the electrode. Those calculations demonstrate that the proposed devices can reach ON/OFF ratios up to 2000 for a running current in the 10-μA range.

As an example, the $T(E)$ spectra and I–V curves for the AND gate are shown in Fig. 3. The device operates like an AND gate only at the negative bias voltage range, where the tunneling current changes by at least a factor of 10 for ON and OFF states. Notice that in those calculations, no surface relaxation was considered to rely on systematic design rules. DFT calculations have demonstrated that our semi-classical design is robust against the known surface relaxation of the Si(100)H surface with dangling bond dimer simply because the output current is measured in the surface gap. It is therefore less sensible to the detail electronic resonance position found in the valence and conduction band energy ranges of the Si(100)H surface. The atomic scale surface configuration of all the others Boolean logic gates is also given in Fig. 4.

As presented in Fig. 5, a simple acetophenone molecule can easily realize the latching effect of the two in and out hydrogen dimer of each of the atomic switches used in Fig. 4. This is because the end phenyl of the acetophenone will adopt the well-known butterfly surface configuration when approaching a DB dimer

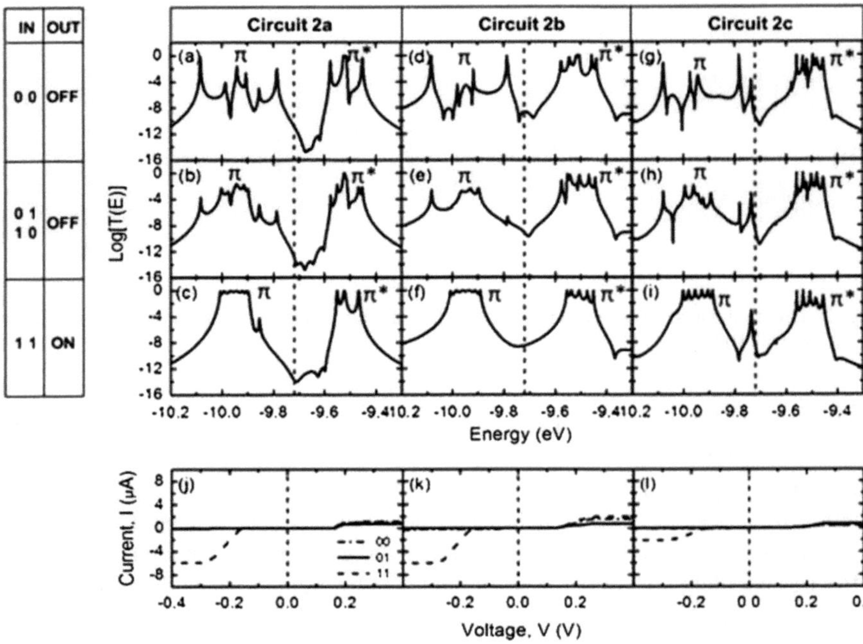

Fig. 3 *T*(*E*) electronic transmission spectra and *I–V* curves for the AND gates in Fig. 4. (3*a–c*) *T*(*E*) spectra for the AND circuit with a central Au nano-island, Fig. 4a; (3*d–f*) *T*(*E*) spectra for the AND gate without the central Au nano-island, Fig. 4b; (3*g–i*) *T*(*E*) spectra for the AND gate interconnected by DB atomic wires and Au nano-pads, Fig. 4c. The corresponding logic input and output of the devices are presented in the *left columns*. The *I–V* curves corresponding to the (*a–c*), (*d–f*) and (*g–i*) spectra are shown in (*j*), (*k*) and (*l*), respectively

atop. When fully chemisorbed, this end phenyl will re-saturate electronically the DB dimer almost as if they were the two corresponding hydrogens, at least in the considered energy range around the Fermi energy (see Chapter "Molecule-Latches in Atomic Scale Surface Logic Gates Constructed on Si(100)H" for a complete discussion about molecule latches).

The interest of this hybrid inorganic–organic approach is that each molecule latch is independent of the others in the circuit like in any Boolean logic circuit as originally proposed by C. Shannon in 1937 at a macroscopic scale [10]. A molecule latch can be switched ON/OFF by an STM tip. The planarization of such a mechanical molecular input is technologically expected. It requires two planar nano-metallic pads more than for the switch design in Fig. 1 because one needs to deliver the required input tunneling current for switching the latch. This is fine for the planarization of this single switch in Fig. 1 whose planar version is device in Fig. 5 . However, for more complex logic circuits than Fig. 1, this is imposing some design constrains, for example, due to the tunneling surface leakage current appearing between the many metallic nano-pads certainly required in this case. For this reason and as a generalization of Chapter "Quantum Hamiltonian Computing (QHC) Logic Gates"

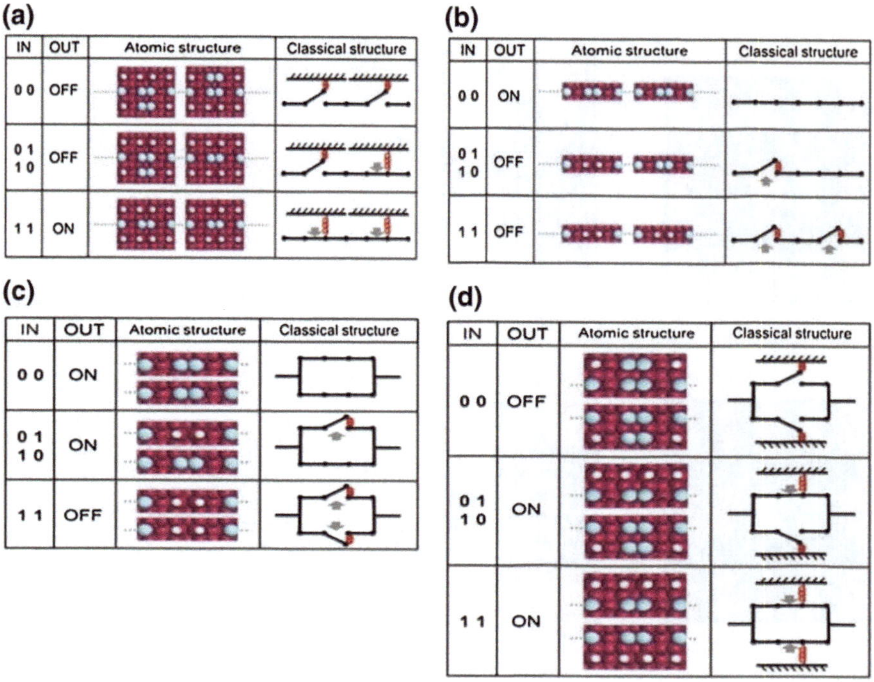

Fig. 4 Truth tables and the atomic structures of the input for **a** AND, **b** NOR, **c** NAND and **d** OR gates. For each atomic scale input configuration, the equivalent electromechanical circuit is shown

QHC circuit design, we are presenting below the design of surface QHC dangling bond DB circuits in a way to minimize the number of molecule latches to be able to use them only once per data input and not to be active on each of the Boolean logical functioning along the circuit.

3 A QHC Dangling Bond Logic Design on Si(100)H

As recalled in Chapter "Quantum Hamiltonian Computing (QHC) Logic Gates", the QHC Boolean logic approach is based on the control of the intrinsic time-dependent evolution of a well-designed quantum system by the logical inputs. The electronic structure of the gate, i.e., the Hamiltonian matrix elements depend on the practical way chosen to encode the logical inputs and to measure the logical outputs of the gate. Very simple three atoms gates have been proposed and constructed experimentally in Chapter "Quantum Hamiltonian Computing (QHC) Logic Gates" with an STM. Starting from this experimental demonstration, the QHC approach can be expended using a quantum graph approach [10]. As recalled in Fig. 6 and starting from the three calculating states (plus the two states of the reading block), one can

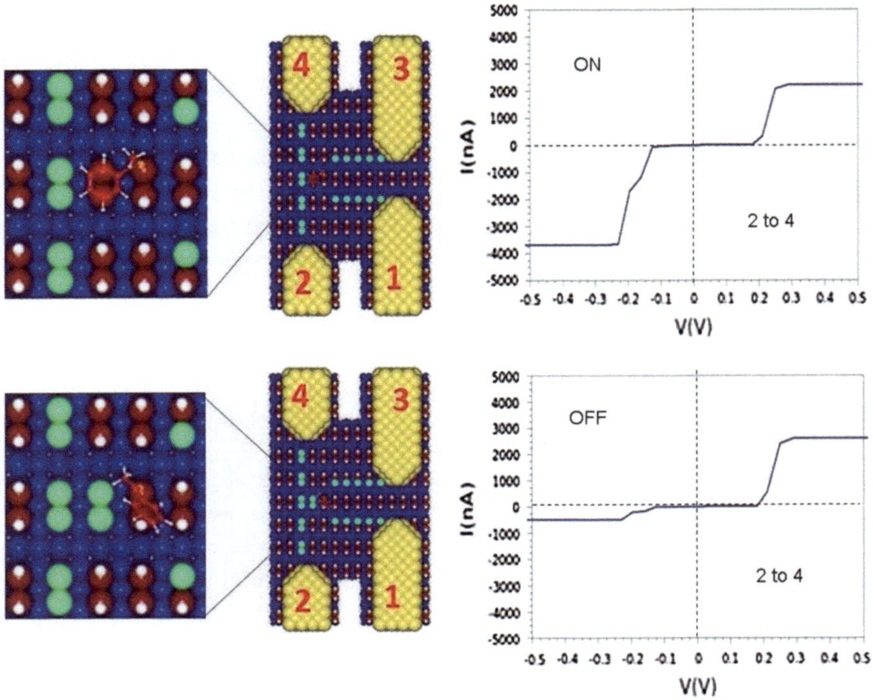

Fig. 5 Simple design of an acetophenone latch triggered by a bias voltage supposed to be applied between Au nano-pads *1* and *3*. The "ON" and "OFF" *I–V* curves were calculated with the N-ESQC technique. The in/out current intensity is measured between electrodes *2* and *4* (see Chapter "Molecule-Latches in Atomic Scale Surface Logic Gates Constructed on Si(100)H" for more details)

generate different quantum graphs which will all fulfilled some QHC Boolean logic functions depending on the symmetry of the graph. In this figure, we have followed the QHC DB 2×1 surface logic gate path (bottom right of Fig. 6). The left part of this QHC graph filiation corresponds to QHC molecules explored first using the starphene molecule [4].

3.1 QHC Surface DB Logic Gates

Using the QHC quantum graphs proposed Fig. 6, we have constructed a 10 × 10 QHC matrix with its 2 × 2 reading head up left block and its 8 × 8 calculating down right block:

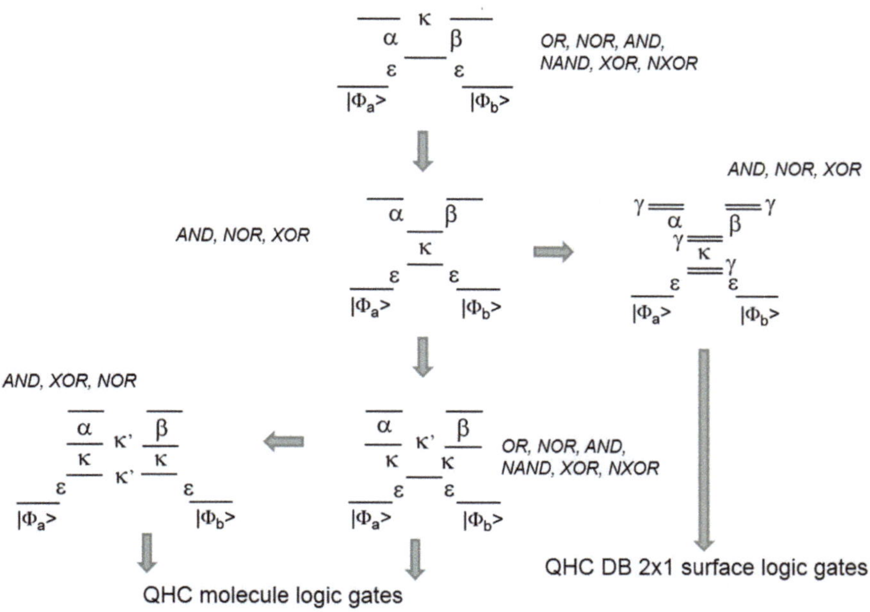

Fig. 6 Starting from Chapter "Quantum Hamiltonian Computing (QHC) Logic Gates" Fig. 4 simple QHC logic gate, a complete series of QHC logic gate quantum graphs is obtained by playing with the $\Delta_{\alpha\beta}$ zero-order term of the calculating matrix characteristic polynomial of the QHC matrix calculating block

$$H(\alpha,\beta)=\left(\begin{array}{cc|cccccccc} E & 0 & \epsilon & \epsilon & 0 & 0 & 0 & 0 & 0 & 0 \\ 0 & E & \epsilon & \epsilon & 0 & 0 & 0 & 0 & 0 & 0 \\ \hline \epsilon & \epsilon & e & d & k & k & 0 & 0 & 0 & 0 \\ \epsilon & \epsilon & d & e & k & k & 0 & 0 & 0 & 0 \\ 0 & 0 & k & k & e & d & \alpha & \alpha & \beta & \beta \\ 0 & 0 & k & k & d & e & \alpha & \alpha & \beta & \beta \\ 0 & 0 & 0 & 0 & \beta & \beta & e & d & 0 & 0 \\ 0 & 0 & 0 & 0 & \beta & \beta & d & e & 0 & 0 \\ 0 & 0 & 0 & 0 & \alpha & \alpha & 0 & 0 & e & d \\ 0 & 0 & 0 & 0 & \alpha & \alpha & 0 & 0 & d & e \end{array}\right)$$

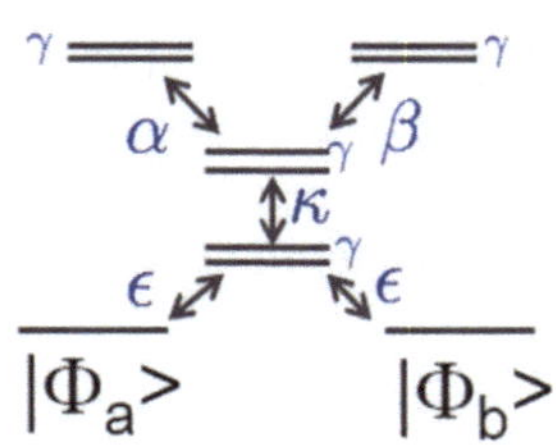

Following Chapter "Quantum Hamiltonian Computing (QHC) Logic Gates" QHC model, each doublet of this QHC graph is supposed to represent a DB dimer of the Si(100)H surface. Because of its symmetry, this QHC matrix is only able to give rise to NOR, XOR and AND Boolean logic gates. This can be demonstrated by determining the conditions for the $\Delta_{\alpha\beta}$ zero-order term of the 8×8 matrix characteristic polynomial to be zero depending on α and β. $\Delta_{\alpha\beta}$ can be written:

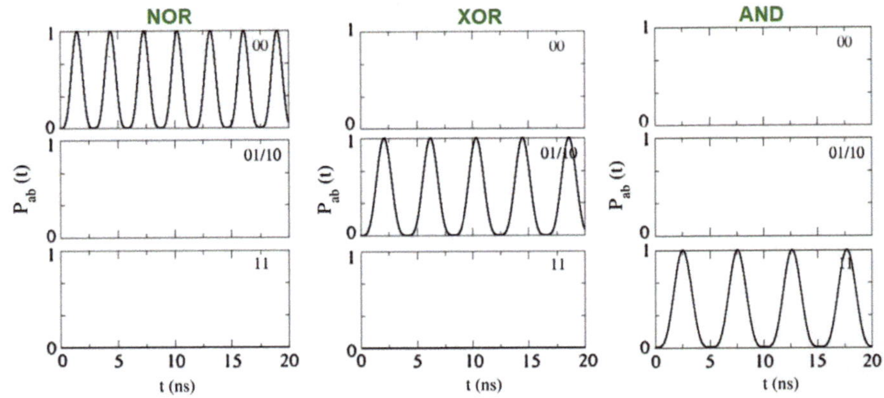

Fig. 7 Characteristic Rabi oscillation between $|\phi a\rangle$ to $|\phi b\rangle$ through the central eight quantum states calculating system. The NOR was obtained for d = 2, κ = 1 and e = (2 × $3^{1/2}$ − 2), the XOR for d = 2, κ = 1 and e = 0 and the AND for d = 2, κ = 1 and e = (2 × $3^{1/2}$ + 2) all in eV

$$\begin{aligned}\Delta_{\alpha,\beta} &= (\alpha^2+\beta^2)(-4e^6+8d^5+4d^2e^4-16d^3e^3+4d^4e^2+8d^5e-4d^6)\\&\quad+e^8-4e^6(k^2+d^2)+e^5(8dk^2)+e^4(4d^2k^2+6d^4)+e^3(-16d^3k^2)\\&\quad+e^2(4d^4k^2-4d^6)+e(8d^5k^2)-4d^6k^2+d^8\end{aligned}$$

After solving this equation for the different α and β logical input values, one can get the 8 × 8 structural matrix term leading to the three possible Boolean logic gates with this matrix. As indicated in Chapter "Quantum Hamiltonian Computing (QHC) Logic Gates", a simple way to measure the Boolean logic response of a QHC gate is to calculate the Heisenberg–Rabi oscillations between the states $|\phi a\rangle$ and $|\phi b\rangle$ of the reading block through the central eight quantum states. This is presented in Fig. 7 confirming the NOR, XOR and AND gates functioning of the above quantum graph.

3.2 A Si(100)H QHC Implementation with Dangling Bond Logical Inputs

Using the symmetry and the DB states available on a Si(100)H surface, the above QHC quantum graph can be easily implanted on this surface by extracting 2 DB from the same dimer on the same row to create the two quantum states required per QHC graph node. This leads to the Si(100)H surface structure presented in Fig. 8 where the logical input encoding has been chosen similar to Chapter "Quantum Hamiltonian Computing (QHC) Logic Gates" by attributing a logical "0" when there are no hydrogen atoms on a DB and a logical "1" when here the two hydrogens are saturating the two Si surface atoms of the same dimer.

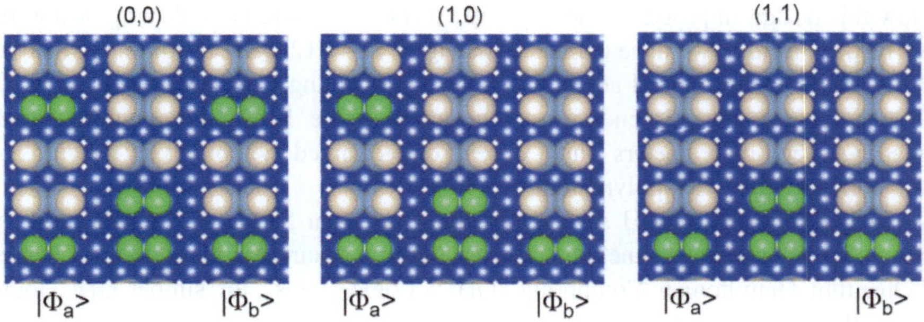

Fig. 8 Surface implantation of the NOR, XOR and AND Boolean logic gates whose simplified QHC matrix is the 10 × 10 matrix given above

But to re-enforce the electronic interactions between the inputs and the core of this QHC gate, a complementary top DB dimer was added to design in Fig. 8. This comes from the fact that to go from the quantum graph to design in Fig. 8, one has also to consider the detail surface atomic scale structure of the Si(100)H surface where the electronic surface coupling between two consecutive SiH dimer rows is rather very weak. The new implementation is presented in Fig. 9. To trouble check

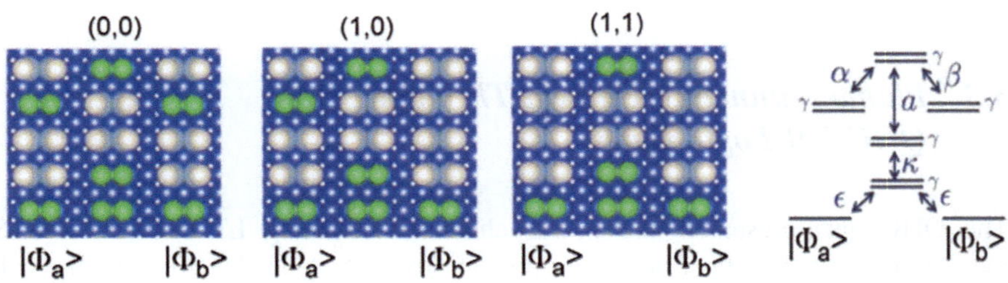

$$H(\alpha,\beta)=\left(\begin{array}{cc|cccccccccc}
E & 0 & \epsilon & \epsilon & 0 & 0 & 0 & 0 & 0 & 0 & 0 & 0\\
0 & E & \epsilon & \epsilon & 0 & 0 & 0 & 0 & 0 & 0 & 0 & 0\\
\hline
\epsilon & \epsilon & e & d & \kappa & \kappa & 0 & 0 & 0 & 0 & 0 & 0\\
\epsilon & \epsilon & d & e & \kappa & \kappa & 0 & 0 & 0 & 0 & 0 & 0\\
0 & 0 & \kappa & \kappa & e & d & a & a & 0 & 0 & 0 & 0\\
0 & 0 & \kappa & \kappa & d & e & a & a & 0 & 0 & 0 & 0\\
0 & 0 & 0 & 0 & a & a & e & d & \alpha & \alpha & \beta & \beta\\
0 & 0 & 0 & 0 & a & a & d & e & \alpha & \alpha & \beta & \beta\\
0 & 0 & 0 & 0 & 0 & 0 & \alpha & \alpha & e & d & 0 & 0\\
0 & 0 & 0 & 0 & 0 & 0 & \alpha & \alpha & d & e & 0 & 0\\
0 & 0 & 0 & 0 & 0 & 0 & \beta & \beta & 0 & 0 & e & d\\
0 & 0 & 0 & 0 & 0 & 0 & \beta & \beta & 0 & 0 & d & e
\end{array}\right)$$

Fig. 9 Surface implantation of the NOR, XOR and AND Boolean logic gates whose simplifed QHC matrix is the 12 × 12 matrix given in this figure. This QHC gate was STM constructed and tunneling spectroscopy characterized [11] showing a nice spectral NOR and OR logical functioning at positive voltage. The existence of an OR functioning not corresponding to the QHC graph theory is given in the text how this design imposed by the Si(100)H surface structure is modifying or not the QHC design rules, we have extracted for Fig. 9 12 × 12 QHC Hamiltonian matrix. This matrix is reproduced above and is also performing like NOR, XOR or AND Boolean logic gates depending on the values of the 10 × 10 calculating block matrix structural parameters which were also determined by looking for the zeros of the corresponding $\Delta_{\alpha\beta}$ polynomial.

In Fig. 9, the Si(100)H surface structure has been STM constructed atom by atom and its scanning tunneling spectrum (STS) measured as presented in Chapter "Quantum Hamiltonian Computing (QHC) Logic Gates" for simple QHC logic gate. By positioning the tip apex on the central row exactly in between the two input dimers, NOR and OR Boolean logical outputs were measured experimentally [11]. This points out that the simple quantum graphs in Fig. 6 are not enough for a fine tuning of the QHC logic gate structures. One reason is that for the graphs shown in Fig. 6, all the inputs and structural parameters are considered to be independent from each other. This is not the case on a surface. This has the consequence to introduce some relations between the parameters controlling the zeros of the $\Delta_{\alpha\beta}$ polynomials leading to the re-apparition of some Boolean truth tables canceled for symmetry reason. This is the case for the Fig. 9 by graph symmetry, the OR cannot be reached. But introducing a logical relation between, for example, the inputs and the structural parameters, the OR is re-introduced as it was demonstrated experimentally [11].

3.3 *Planar Tunneling Current Through QHC DB Logic Gates*

The QHC gate presented above was characterized using LT-UHV-STM STS measurements in a vertical like measurement procedure [11]. To explore what will happen for a complete surface planar implantation, N-ESQC transport calculations were performed for $N = 2$. Very simple DB semi-infinite wires were used for interconnects keeping first the simple H dimer extraction logical input encoding. Then, acetophenone molecule latches already used above for the semi-classical circuits in Fig. 5 have been tested.

The infinite dangling bond interconnects used here have a very important role to measure the secular Heisenberg–Rabi oscillations frequency capture by the reading head. Those time-dependent oscillations are normally too fast to be recorded experimentally in real time. However, the tunneling current intensity through the central measuring state of the calculating block is proportional to the square of this frequency as soon as the corresponding $T(E)$ is not reaching unity [8]. Notice that metallic nano-pads will play a better role in this quantum to classical measurements because of their more regular electronic density of states near their Fermi energy as compared to dangling bond interconnects. This will be explored in Chapter

"Complex Atomic-Scale Surface Electronic Circuit's Simulator Including the Pads and the Supporting Surface" of this book.

The first set of calculations was performed to certify the interconnection band structure of the semi-infinite DB wires chosen parallel to the dimer rows with no relaxation. As presented above, the active part of the QHC device is built up starting from a perpendicular to the dimer rows short DB wire constructed with 3 DB dimers. To avoid any corner destructive interference effects, we have also added 2 DB dimers leading to a 5 DB dimer atomic wire as presented in Fig. 10. Transport calculations were performed using 2-ESQC in a full valence basis set to describe both the supporting slab and the surface electronic structure. Five Si layers below the surface were necessary to stabilize the surface band structure, the last bottom Si layer being full hydrogen saturated for the ESQC calculations. As showed in Fig. 10, the calculated *T(E)* is presenting large values in the valence band and conduction band energy range of the Si(100)H surface. A moderate *T(E)* is obtained in the energy range of the conduction band of the two semi-infinite DB wires, and the two resonance sets are corresponding to the 5 very short perpendicular to the rows DB dimer wire. The Fermi level was positioned in the gap between those resonances massifs.

In this simple atomic scale interconnection circuit shown in Fig. 10, one can easily position the QHC DB logic circuit presented in Fig. 9. The 2-ESQC calculations of the corresponding *T(E)* spectra were performed at each step of the construction using the same semiempirical parameters than in Fig. 5. At the energy location of the central conduction band and according to Fig. 11, the largest *T(E)* is for the (1,1) logical input configuration where *T(E)* is almost reaching unity for some energy values. The (1,0) and (0,0) logical input configurations are showing up a much lower *T(E)*. This is due to the energy repulsion effect as compared to the

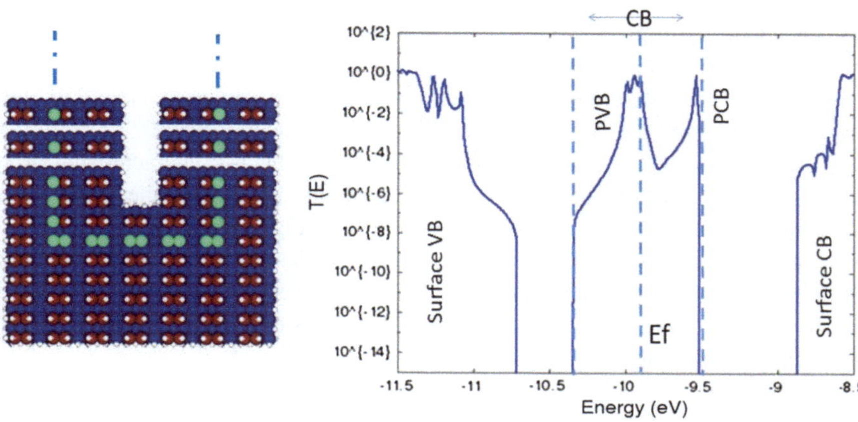

Fig. 10 Calculated *T(E)* transmission coefficient through the central five DB dimers. The energy location of the surface valence band (Surface VB), the surface conduction band (Surface CB) and the semi-infinite conduction band (CB) is clearly visible. *Green* DB Si atoms, *white* H surface atoms, *red* Si surface atoms and *blue* bulk Si atoms

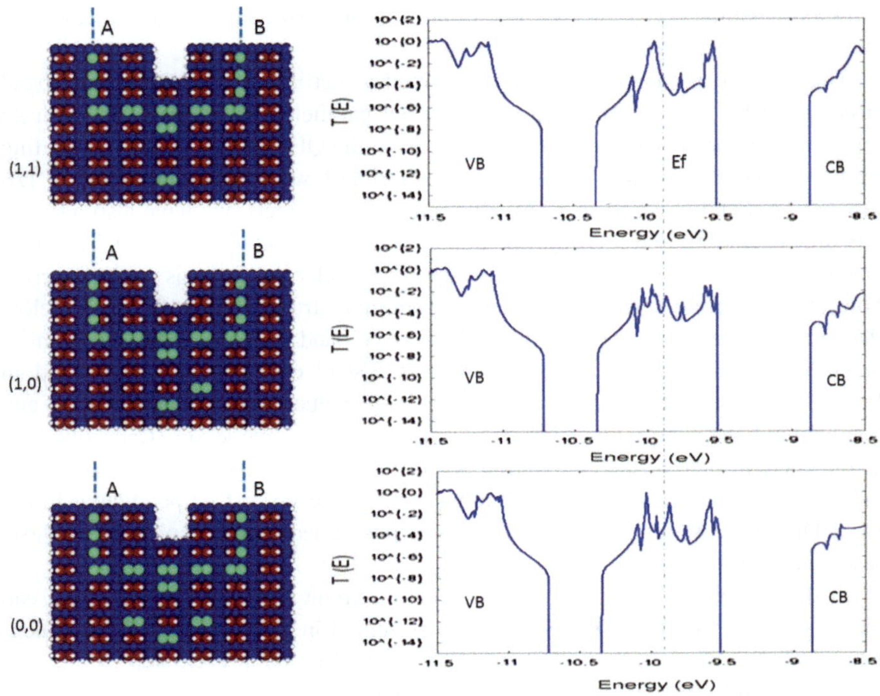

Fig. 11 Calculated *T(E)* through the QHC DB Boolean logic gate constructed above. The input encoding was chosen for the (1,1) to be the full hydrogenated input dimers and the (0,0) after the extraction of 4 H from the surface, two by logical input. The *T(E)* was calculated through the full Si(100)H surface structure from semi-infinite DB wire *A* to the semi-infinite wire *B*

(1,1) configuration because of the through-surface electronic interactions between the new added DB dimers. Notice also that the small destructive inference showing up for (1,1) is not destroyed for (1,0) and (0,0). This destructive interference is introduced by the DB dimer added laterally and in the center of the 5 DB wire perpendicular to the dimer rows on this Si(100)H surface.

To trouble check the obtained logical truth table, the *I–V* characteristics were calculated for each of the logical input configuration using the generalized Landauer formula. The *I–V* curves are presented in Fig. 12 and confirm that in a planar configuration with DB wire interconnects, the designed QHC logic gate is an AND gate, one of the possible QHC gate which can be constructed with this surface implementation. Notice the voltage stability margin is quite large for $V < -0.2$ V, indicating that the surface relaxation of this gate will lead to the same result.

The transport calculations through the same DB logic gate design but with now one acetophenone latching action per logical input are presented in Fig. 13. It was calculated using also the 2-ESQC technique with the molecule latch conformation optimized using the ASED+ molecular mechanics semiempirical approximation. A nice AND gate is also obtained with even a better zero current output for the (0,0)

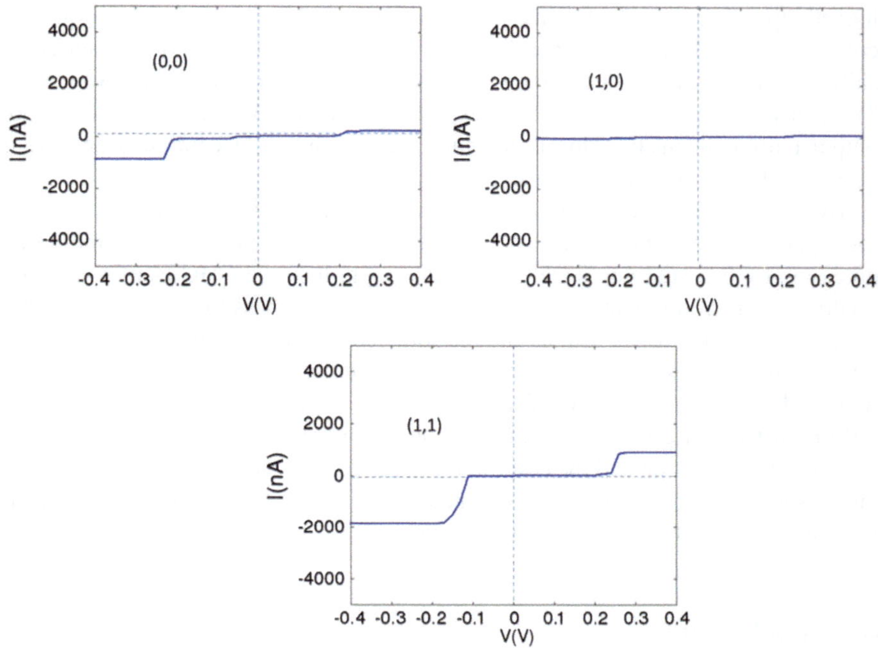

Fig. 12 Calculated *I–V* characteristics for Fig. 9 QHC DB logic circuit. This was calculated starting from *T*(*E*) spectra in Fig. 11 using the generalized Landauer formula. An AND gate behavior is clearly showing up which can also be improved using molecular latches inputs (see below)

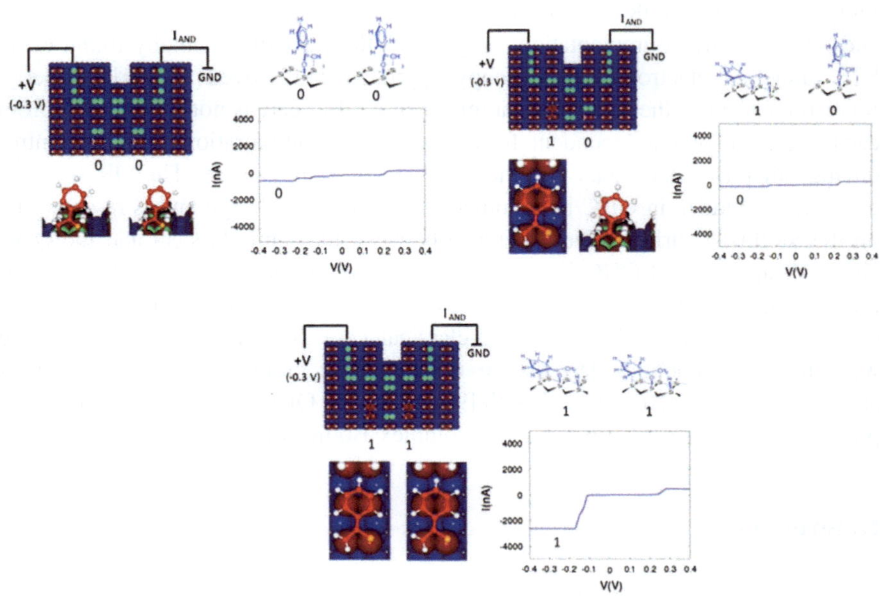

Fig. 13 QHC AND logic gate *I–V* characteristics using one acetophenone molecule latch per logical input. The AND logical input is clearly obtained with a large voltage stability range

logical input that is for the two molecule latches in their vertical conformation. This comes from the fact that the bonding of the acetophenone pivot on the surface is shifting in energy the states of the corresponding Si as compared to an H bonding. This is also smoothing some resonances leading to a much lower (0,0) configuration output current intensity. This is not the case for the (1,1) configuration current because all the input dimers are saturated in this case.

By changing the DB configuration along the 5 DB central atomic, other QHC logic gate Boolean truth tables can also be constructed following the same procedure, i.e., by first studying a QHC quantum graph; then the QHC Hamiltonian and finally the surface-implanted T(E) spectra. Notice that to pass from one logic gate to another, it is not possible just to change some electronic couplings starting from a given atomic scale surface construction because of the very specific structure of the Si(100)H surface. A full new DB dimers configuration has to be defined on the Si (100)H surface for each considered QHC quantum graph. This comes from the fact that the level repulsion and the control of the destructive interference effects at work on the surface for QHC must be fine-tuned in regard also to the bias voltage used to measure the output logical status of the gate.

4 Conclusion

DB Boolean logic gate atomic scale circuits have been designed on a Si(100)H surface and the use of molecule latch inputs also first explored. For semi-classical design and for QHC design, all the four known symmetric 2-input/1-output Boolean logic gates (AND, NOR, NAND and OR) have been found. For QHC DB logic gate and going to surface implantation, only the NOR was experimentally found using STS tunneling electronic spectroscopy approach and proved to work properly. Since on a surface, the structural parameter of a QHC gate is not independent from each other and not independent from some input configurations, this is reintroducing other Boolean gates that the NOR in the present case. The OR was, for example, forbidden in QHC by symmetry in the present design and is re-emerging on the Si(100)H surface. The difference between the semi-classical and the QHC circuit designs is that QHC is minimizing the wiring inside the gate. As a consequence, the measuring tunneling current has not to pass all along the circuit in the QHC case as compared to the semi-classical case. This is a way to avoid the well-known exponential decrease in the tunneling current intensity through at atomic wire as a function of its length [9] making the QHC approach certainly more practical for the construction of more complex Boolean logic gates.

References

1. Joachim, C., Renaud, N., Hliwa, M.: The different designs of molecule logic gates. Adv. Mater. **24**, 312 (2012)

2. Magoga, M., Joachim, C.: Towards circuitry in a tunnel barrier. Phy. Rev. B, **59**, 16011 (1999)
3. Ami, S., Hliwa, M., Joachim, C.: Molecular “OR” and “AND” logic gates integrated in a single molecule. Chem. Phys. Lett. **367**, 662 (2003)
4. Soe, W.H., Manzano, X., Renaud, N., De Mandoza, P., De Sarkar, A., Ample, F., Hliwa, M., Echevaren, A.M., Chandrasekhar, N., Joachim, C.: A single molecule NOR gate with Au atom inputs. Phys. Rev. B, 155443 (2011)
5. Joachim, C.: Probing intramolecular circuit laws. Nat. Nanotechnol. **7**, 620 (2012)
6. Joachim, C.: Single molecule Boolean logic gates. In: Molecular NanoArchitectronics. Springer Series: Advances in Atom and Single Molecule Machines, vol. IX (2016, in press)
7. Kawai, H., Ample, F., Qing, W., Yeo, Y.K., Saeys, M., Joachim, C.: Dangling bond logic gates on the Si(100)-(2x1)-H surface. J. Phys. Cond. Mat. **24**, 095011 (2012)
8. Faizy, O., Dridi, G., Joachim, C.: Parallel quantum circuit in a tunnel junction. Sci. Rep. **6**, 30198 (2016)
9. Kawai, H., Yeo, Y.K., Saeys, M., Joachim, C.: Conductance decay of surface hydrogen tunnelling junction fabricated along a Si(100)-(2x1)H atomic wire. Phys. Rev. B **81**, 195316 (2010)
10. Shannon, C.E.: Asymbolic analysis of the Relay and switching circuits. Master thesis, Bell Labs Archives (1937)
11. Kolmer, M., Zuzak, R., Godlewski, S., Szymonski, M., Dridi, G., Joachim, C.: Realization of a quantum Hamiltonian computing Boolean logic gate on the Si(001):H surface. Nanoscale **7**, 12325 (2015)

Molecule-Latches in Atomic Scale Surface Logic Gates Constructed on Si(100)H

Francisco Ample, Francesca Moresco and Christian Joachim

Abstract To control the logical inputs in a surface atomic scale circuit, we introduce the concept of molecule-latch. The switching of certain molecules on a surface can be used to control the conductance of a very simple atomic scale interferometer that can be constructed atom by atom on an Si(100)H surface by STM atomic manipulation. Among the possible surface switching effects, such as surface rotation, push–pull effect or flip-flop switching, we explored the latter switching effect in detail starting from the acetophenone molecule, which can be switched on the Si(100) surface by an inelastic tunnelling current effect. Since this molecule does not offer a true bistable character on the surface, longer-arm acetophenone-like molecules are introduced, keeping the end phenyl unchanged. Among these molecules, 4-acetylalkylphenyl presents the best bistable character on Si(100)H and can perfectly control the conductance of a Si(100)H atomic scale electronic surface interferometer.

F. Ample
Institute of Materials Research and Engineering, 2 Fusionopolis Way, Innovis, #08-03, Singapore 138634, Singapore

F. Moresco
Max Bergmann Center of Biomaterials, Center for Advancing Electronics Dresden, Institute for Materials Science, TU Dresden, 01062 Dresden, Germany

C. Joachim (✉)
GNS & MANA Satellite, CEMES-CNRS, 29 rue J. Marvig, 31055 Toulouse Cedex, France
e-mail: joachim@cemes.fr

C. Joachim
International Center for Materials Nanoarchitectonics (WPI-MANA), National Institute for Materials Science (NIMS), 1-1 Namiki, Tsukuba, Ibaraki 305-0044, Japan

M. Kolmer and C. Joachim (eds.), *On-Surface Atomic Wires and Logic Gates*, Advances in Atom and Single Molecule Machines,
DOI 10.1007/978-3-319-51847-3_10

1 Introduction

A molecule-latch is a single molecular switch that can maintain its ON/OFF status after being activated by an external stimulus. Whatever conformation the molecule has after its excitation, it will persist until the molecule is actuated again. This is an important device to design logical inputs for classical and semi-classical atomic scale circuits, but certainly also for the quantum Hamiltonian computing (QHC) approach (see, for example, Chapter "The Design of a Surface Atomic Scale Logic Gate with Molecular Latch Inputs"). In this last case, the concept of a latch is crucial since after changing the logical status of one input on a QHC circuit, this new logical input status must be preserved during a certain time interval even if the stimulus used for this change is interrupted. This time interval is necessary for the tunnelling current through the QHC output reading block to build up that is to be stabilized. Ideally, the ON and the OFF logical status of a molecule-latch must be equivalently accessible by the controlling stimulus coming from the macroscopic world and maybe one day come from the conversion of quantum information in a mechanical action.

For the design of atomic scale logical circuits, we first considered a molecule-latch based on the acetophenone molecule, whose end phenyl is known to adsorb on a dangling bond (DB) dimer of the Si(100) surface [1]. As presented in Fig. 1, an acetophenone molecule can have two stable conformations on Si(100). It was then experimentally demonstrated that acetophenone adsorbed on the Si(100) surface can switch from one to the other conformation by passing a tunnelling current through it and then by increasing the bias voltage [2]. This opens the quest for using such a switching phenomenon to design a reversible molecule-latch and also to improve the latching performances of a series of acetophenone-like molecules as described below.

As presented in Fig. 2, three types of planar molecule-latches can be designed on a surface by using different molecular switching effects: (1) a flip-flop, (2) a

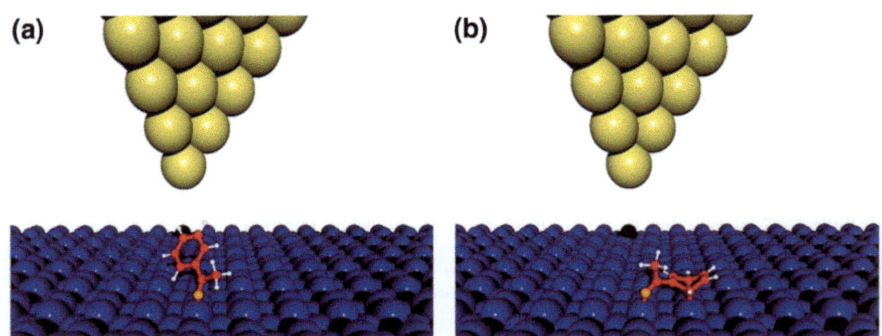

Fig. 1 The two stable conformations of an acetophenone molecule chemisorbed by its molecular pivot on an Si(100) surface dangling bond. One can pass from one conformation to the other by pulsing the tunnelling current passing through this molecule

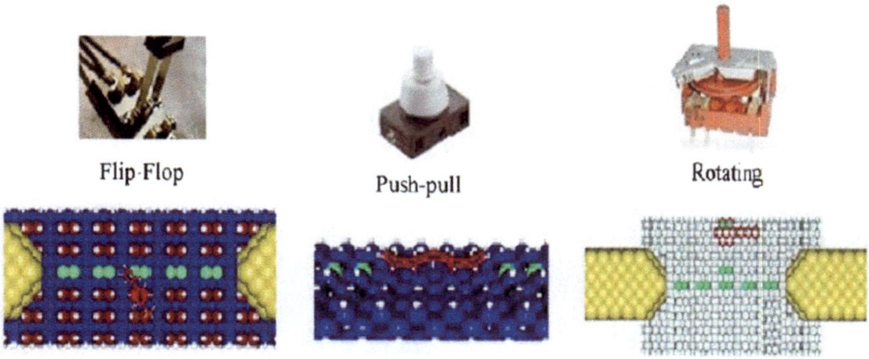

Fig. 2 The three different possible identified molecule-latches and their Si(100)H possible surface implantation. Only the rotating switch had not yet beneficiated from detail *I*–*V* characteristics N-ESQC calculations because of the relatively small energy barrier of rotation while adsorbed on the Ge(0001)H surface

push–pull or (3) a surface rotating switch. Each of them shows a peculiar surface molecule switch configuration as presented in Fig. 2 and leads to a specific atomic scale circuit surface implantation to construct an atomic scale molecular switch as presented in detail in this chapter. After this detail discussion, we will focus in the following on what presently seems to be the most promising switch, i.e., the flip-flop molecule. We will present a semi-classical model to describe how such flip-flop can be driven by inelastic tunnelling current effects with the tip apex of a scanning tunnelling microscope (STM) pointing on a specific location of this molecule when adsorbed on an Si(100)H surface. Then, we will present a series of flip-flop molecules derived from the acetophenone.

2 The Different Molecule-Latches on a Semi-conductor Surface

2.1 A Flip-Flop Molecule-Latch

As presented in Fig. 1, the first example of possible molecule-latch is based on an acetophenone molecule as switching element. As presented in Fig. 3, the two stable conformations of this molecule on an Si(100) (2×1) surface can be determined following the minimum energy switching path on its ground-state potential energy using an ASED+ NEB reaction path optimizer as described in Sect. 4.

As presented in Fig. 4, a simple planar circuit including a molecule-latch can be designed using the flip-flop switching of a single acetophenone molecule adsorbed on Si(100) (2×1). The design is based on molecule–DB bond by the end phenyl in a butterfly-like configuration and then on its saturation by switching the underneath

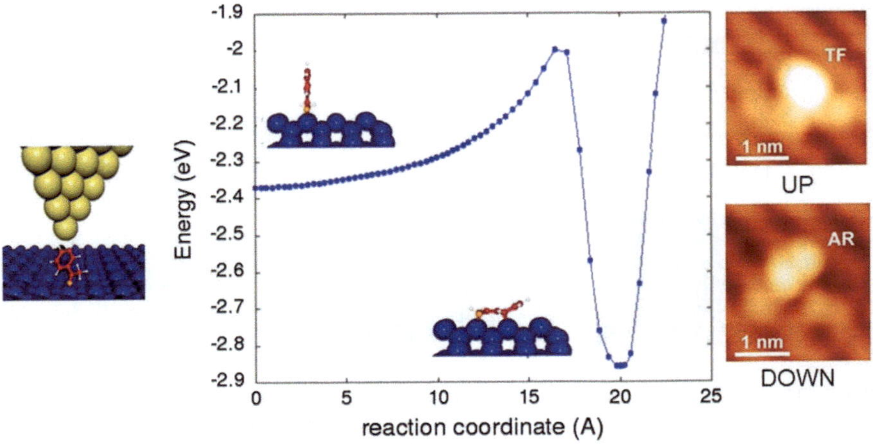

Fig. 3 ASED+ optimized surface conformation of an acetophenone molecule on the Si(100) (2×1) surface with a first estimation of the reaction path between its two surface conformations. The UP and DOWN conformations LT-UHV-STM images were first obtained by UCL, London [2], and were very difficult to reproduce because of the difficulties in safely adsorbing the acetophenone molecules on Si(100)H without triggering side surface chemical reaction which is locally deforming the Si(100) (2×1) surface reconstruction. This first ASED+ calculation gives a switching energy barrier height of 0.4 eV between the 2 conformations. But the latching well is very pronounced (0.5 eV)

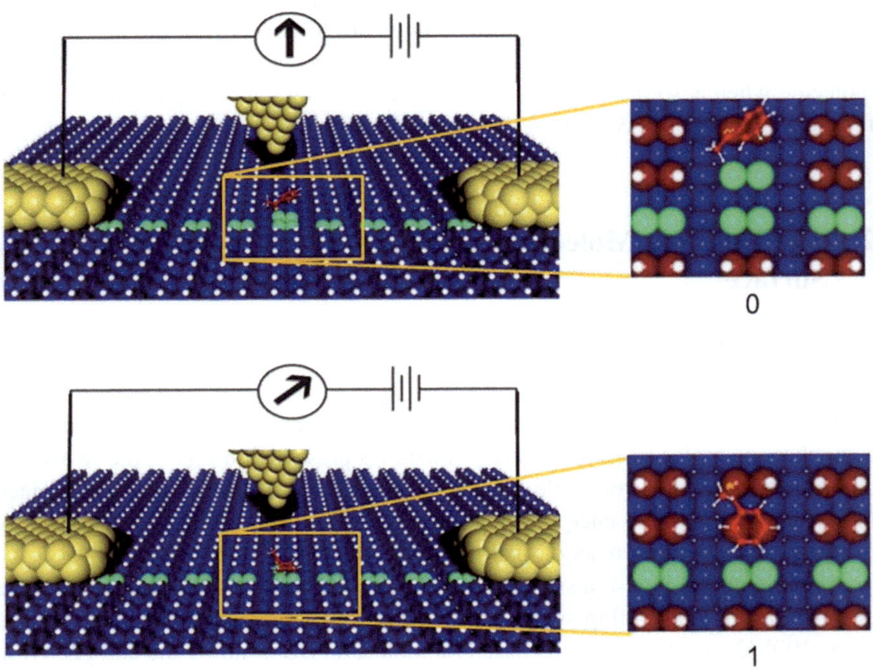

Fig. 4 The surface atomic scale structure of the acetophenone flip-flop molecule-latch with its two input–output nano-pads and the STM tip apex driving electronically the switching of the latch. The *two right panels* give a top view of the OFF ("0") and ON ("1") configurations of the latch

DB dimer, which was constructed on purpose before adsorbing the molecule. This complementary underneath DB dimer constitutes the atomic scale electronic interferometer part of the device. Also in a butterfly-like conformation and on this DB dimer, the chemisorption of the end acetophenone phenyl results in a chemical saturation of this DB dimer.

The *T*(*E*) spectra and the *I–V* curves of this molecule-latch were calculated using the $N = 2$ elastic scattering quantum chemistry (ESQC) calculation technique including the Si(100)H surface leakage current and using only a slab of 5 Si layer below the surface. As presented in Fig. 5, a very elegant atomic scale quantum interferometer results where the lateral dimer to the DB atomic wire plays the role of a lateral branch with a different Bloch wave phase shift as compared to the principal DB wire branch. As a consequence, the tunnelling current through this wire with the acetophenone in the up ("OFF") states is very small in intensity as compared to the "ON" acetophenone conformation where its phenyl is butterfly-adsorbed on the Si(100)H surface dangling bond. In this design, the tip apex of the STM is used to trigger the switching event. There is no need for mechanical push on the molecule to ensure the ON/OFF switching. Then, when the acetophenone is in its either UP or DOWN conformation, there is no need to

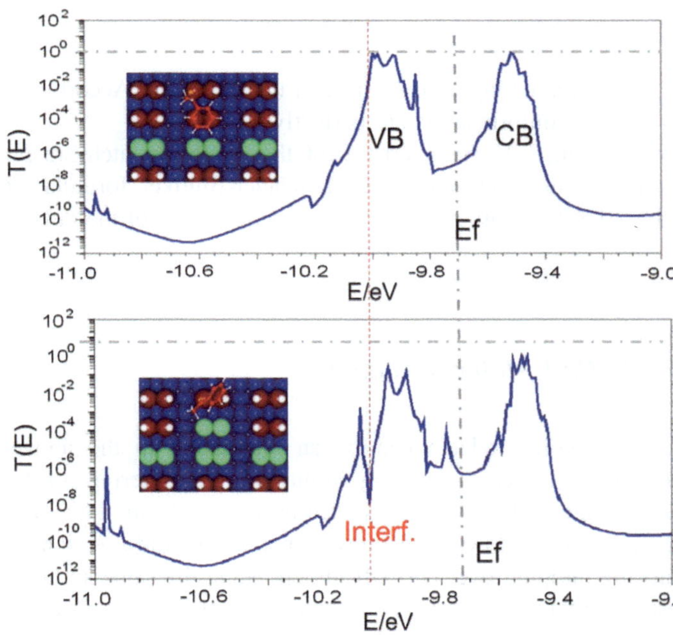

Fig. 5 2-ESQC calculated *T*(*E*) spectra for the ON and OFF configuration of the molecule-latch in Fig. 4 including the metallic contact nano-pads, the Si(100)H surface and the central atomic scale interferometer. In the *T*(*E*) OFF state (*bottom panel*), the destructive interference of the surface dangling bond is clearly visible and is cancelled when the end phenyl of the acetophenone is bond to this dangling bond in its stable butterfly-like surface conformation

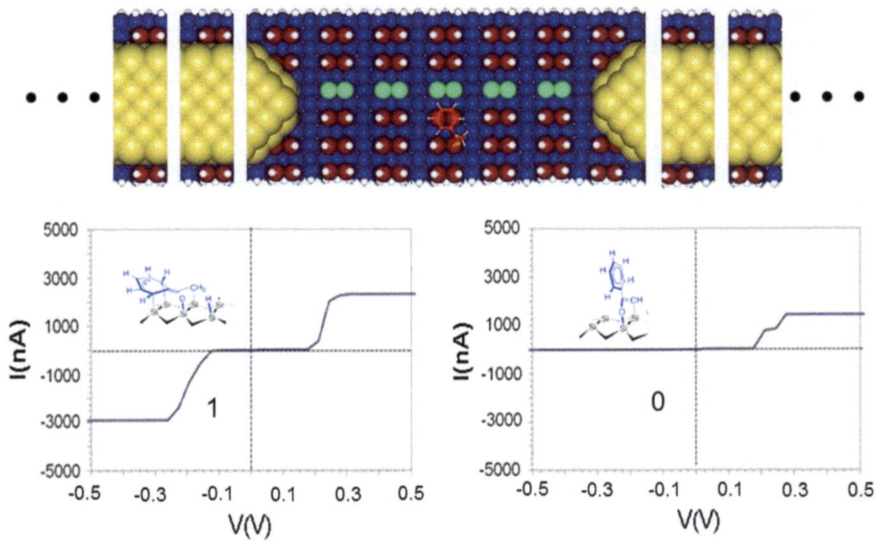

Fig. 6 *I–V* characteristics of the acetophenone molecule-latch. The "ON" and "OFF" conformations of acetophenone are controlling the atomic scale interferometer, leading the tunnelling current through this device to be as large as 3 μA in the "ON" state and a few nA in the "OFF" state

continue to pass a tunnelling current through the molecule. According to computation, both surface conformations are perfectly stable.

The corresponding *I–V* characteristics of this molecule-latch are presented in Fig. 6, calculated using the generalized Landauer-Buttiker formula after having calculated the 2-ESQC multi-channels transmission matrix of the device presented in Fig. 4.

2.2 A Push–Pull Molecule-Latch

A push–pull-like molecule-latch can be an alternative to the above-presented flip-flop molecule and opens a new way of mastering the "spring constant" for the conformation change of the molecular switch inside a molecule-latch. For example, a pentacene molecule is known to chemisorb in a planar conformation on a Si(100) (2×1) surface by jumping over two DB dimers along the same surface row. Here and as presented in Fig. 7, a molecular switching event was observed with the up-down motion of the central part of this molecule on the Si(100)H surface [3].

As presented in Fig. 8, a simple molecule-latch switch can be designed using this push–pull characteristics, keeping the two end phenyls in a butterfly configuration along the Si(100)H dimer row as the stable pinning points of the molecule [3]. Inserting 2 other DB dimers and forming again a surface atomic scale

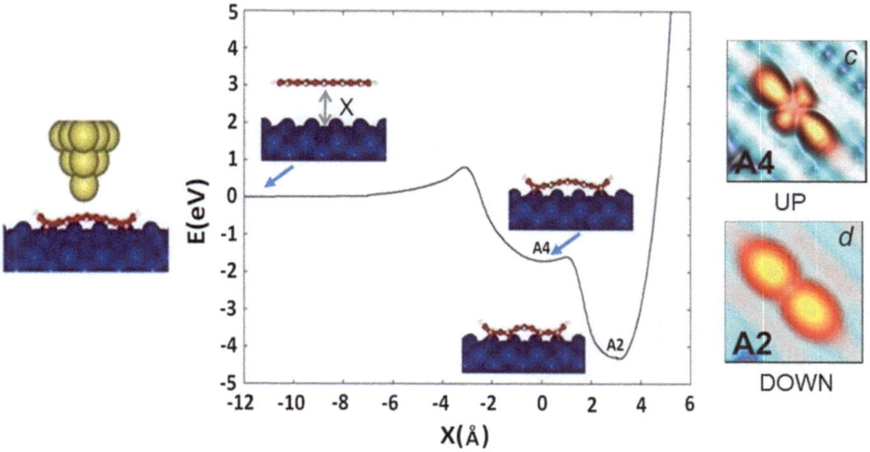

Fig. 7 The minimum switching energy for a pentacene molecule adsorbed in a "double" butterfly-like conformation on the Si(100) (2×1) surface with its two UP and DOWN surface-identified conformations and their corresponding experimental LT-UHV-STM images. Here, there is one metastable A4 and one very stable A2 surface conformation

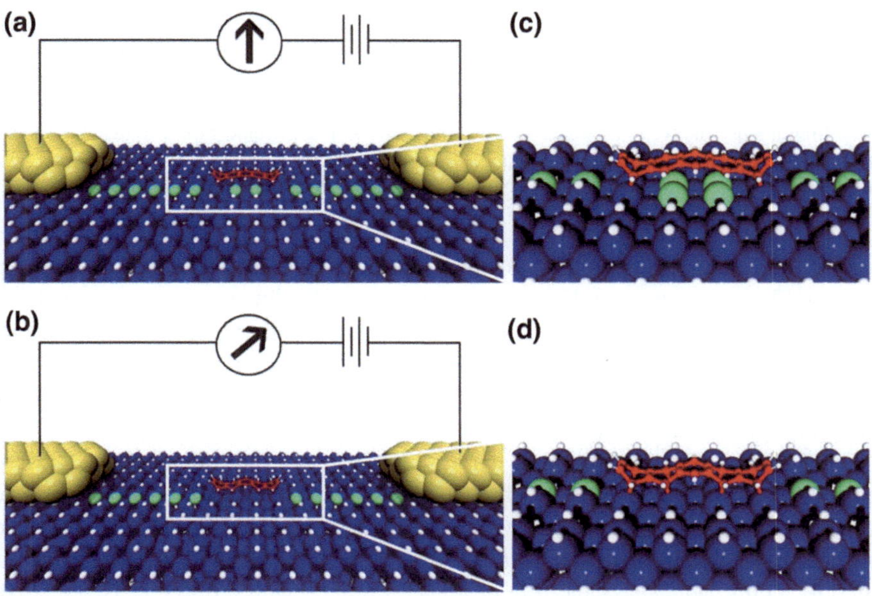

Fig. 8 Surface atomic scale structure of the pentacene push–pull molecule-latch with its two input–output nano-pads. The STM tip apex is supposed to electronically drive the switching of the latch as in the design in Fig. 4. The *two right panels* show a *top view* of the OFF (**c** for "0") and ON (**d** for "1") configurations of the latch. Experimentally, the OFF to ON switching can be easily triggered by an inelastic tunnelling current effect since the OFF state is a metastable state (see conformation A4 Fig. 7 above). From the ON state (see conformation A2 Fig. 7 above), it is very difficult to return to the OFF state because the required *pi** state is in the Si(100) (2×1) band gap and therefore is electronically inaccessible. Only a mechanical push by the STM tip apex end can restore the OFF state A4 surface conformation [3]

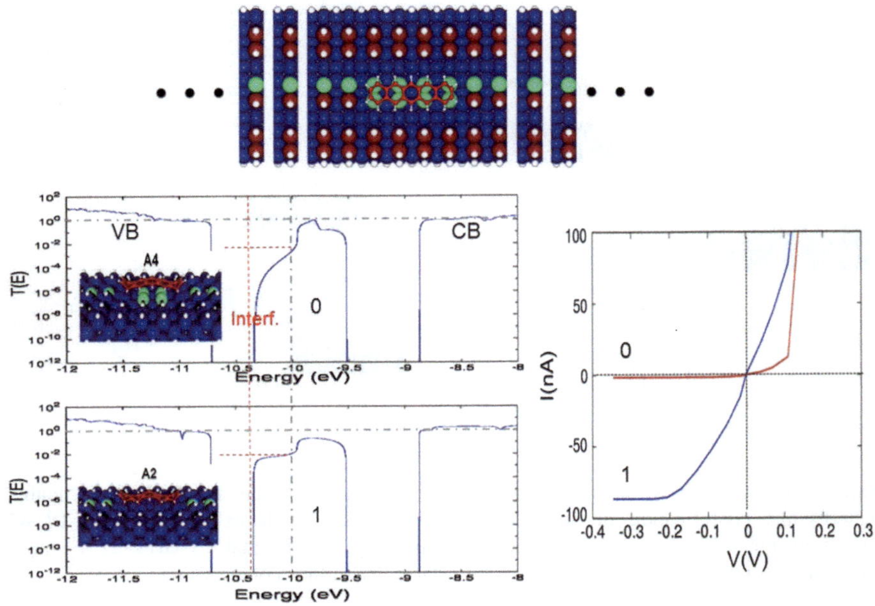

Fig. 9 *I–V* characteristics of the pentacene molecule-latch in Fig. 8 with its corresponding *T*(*E*) spectra. The pentacene ON ("1") and OFF ("0") conformations are controlling the atomic scale interferometer, leading the tunnelling current through this device to be around 90 nA in the "ON" state and still a few nA in the "OFF" state. The binding of the pentacene along the dimer row is shifting the destructive interference in the surface Si(100)H electronic gap. But this interference is still active on the central conduction band. Notice that here we have not relaxed the atomic wire which would have also resulted in an opening of a gap in this central conduction band

interferometer underneath the molecule, a destructive interference effect shows up in the OFF state because in this case those two new DB dimers are not saturated and only weakly coupled to the DB atomic wires. While switching down the pentacene, those two DB dimers are saturated and the tunnelling current intensity increases through this device as presented in Fig. 9 because of the very large electronic coupling through the surface.

The *T*(*E*) spectra and the *I–V* curves of such pentacene molecule-latch were calculated using the 2-ESQC technique including the Si(100)H surface leakage current using only a slab of 5 Si layer below the surface. As presented in Fig. 9, the atomic scale quantum interferometer effect is preserved and the DB dimer lateral to the DB atomic wire plays the role of a lateral branch with a different Bloch wave phase shift as compared to the principal DB branch. The tunnelling current through this wire with the pentacene in an up ("OFF") state is very small in intensity as compared to the "ON" pentacene conformation state. Notice that due to the necessity to bind the pentacene along the dangling bond dimer, the two end phenyls of the pentacene are saturating each a dangling bond dimer leading to overall ON and OFF tunnelling current intensities much larger than for Fig. 4 push–pull molecule-latch. We have also designed an atomic scale interferometer with the same pentacene switching effect but perpendicular to the dimer row. In this case, however, the molecule-latch ON/OFF ratio was much lower as compared to Fig. 8 surface configuration.

2.3 A Rotating Molecule-Latch

A molecule-latch rotating on the surface can be designed using the surface properties of a starphene Y molecule anchored at one end on a DB dimer on the Ge(001)H surface. It can rotate by a simple bias voltage pulse as presented in Fig. 10 [4]. When optimized, such a phenomenon can also be used for a planar rotating latching effect bringing a phenyl over a dangling bond dimer but without butterfly-like adsorption of the end phenyl due to the large lateral surface extension of the Y molecule. To understand the rotating latch effect, we should first understand the surface rotation of a molecule generally physisorbed on a surface by inelastic tunnelling effect, which can, for example, drive a single metal atom [5]. If working on a passivated semi-conductor surface, this will lead to a very useful effect for complexing a single

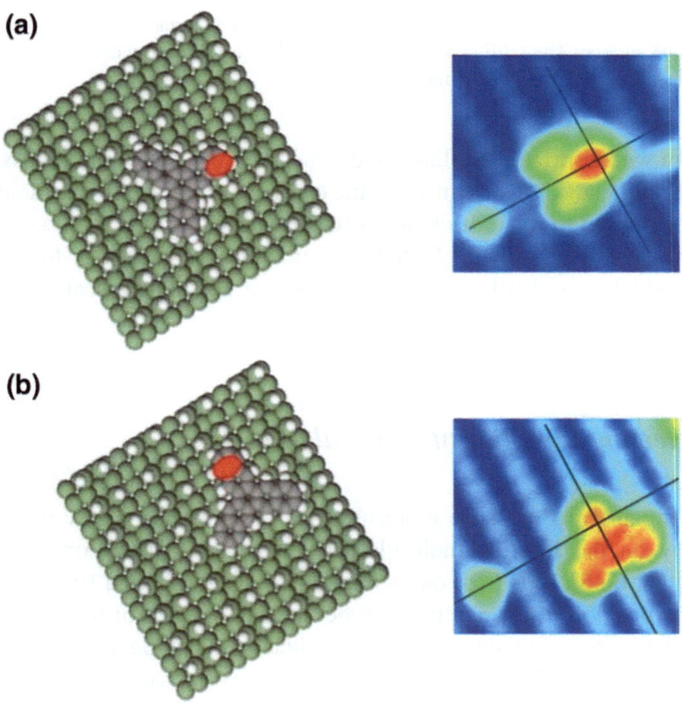

Fig. 10 Induced STM inelastic rotation of a starphene Y molecule on a Ge(001)H surface while one of its end phenyls is bonded to a dandling bond dimer [4]

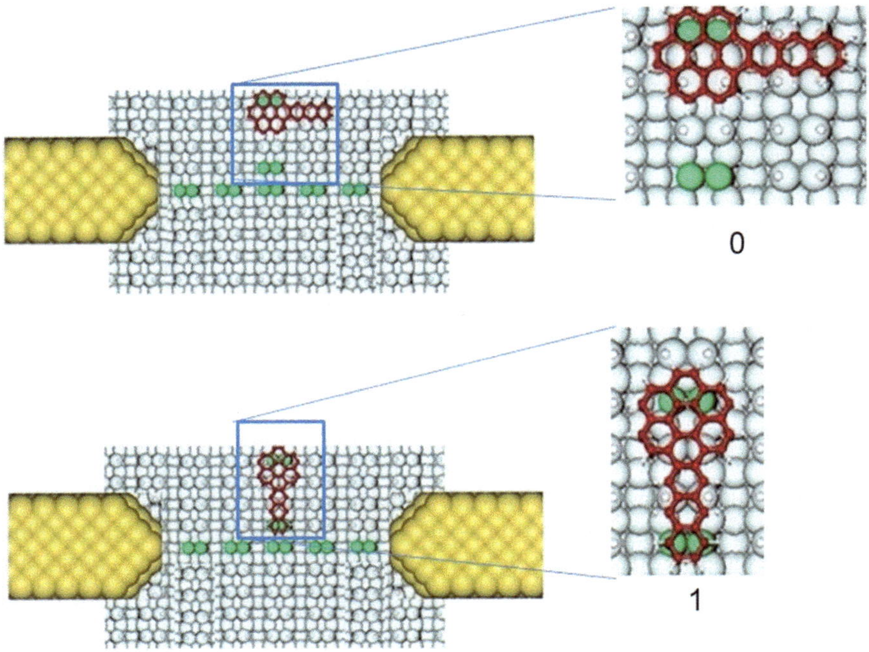

Fig. 11 OFF ("0") and ON ("1") conformations of a possible surface rotating molecule-latch on the Ge(001)H surface. The OFF conformation is based on a DB destructive interference effect as presented in Fig. 4 for the flip-flop molecule-latch

metal atom with a surface dangling bond dimer in a reversible manner. As presented in Fig. 11, this leads us to design a specific coronene–anthracene-like molecule-latch which is supposed to rotate back and forth between 2 surface positions on a passivated semi-conductor surface. 2-ESQC calculations are underway for this rotating molecule which can also lead to the control of a DB atomic scale interferometer (Fig. 11).

2.4 Selecting an efficient Molecule-Latch

After having analyzed possible molecule-latch designs (Fig. 2), we will now focus specifically on the flip-flop molecule-latch of Fig. 4. The rationale behind this choice is that on Si(100)H [or on Ge(100)], the dangling bond surface circuit will perturb the position of the π^* states of the molecule switch of the latch. However, those π^* states are essential to control the switching effect. Separating the switching part (the pivot) from the contacting part is necessary to master the switching effect on the dangling bond dimer. This is exactly what a flip-flop molecular switch is doing. Certainly, the following results can be extended to the other molecule-latch cases. Notice also that to control a surface atomic scale interferometer as designed above, it seems better to have an up-down motion of the end phenyl than a surface rotation to avoid any through surface tunneling leakage current.

3 Inelastic Tunnelling Current Effect acting on an Acetophenone Switch

The latching effect produced by the acetophenone conformation change on the dangling bond device of Fig. 4 is supposed to be triggered by an inelastic tunnel current when passing a tunnelling current only through the acetophenone molecule using the tip of an STM. We have used a semi-classical Ehrenfest approach to describe this inelastic switching effect. In our approach, the electron transfer process through the molecule is described quantum mechanically and the molecule switching mechanics in semi-classical approach [6]. For the inelastic vibronic coupling, we have used a tunnelling current-dependent inelastic force $F_{\text{in}}(I, z)$ developed in the past to understand semi-classically the tunnelling heating of a simple Xe atom in its atomic switch configuration in a STM tunnel junction and on a metal surface [6]:

$$F_{\text{in}}(I,z) = -0.33\gamma_{\text{in}}\tau\frac{I(z)}{e}\frac{\pi\hbar}{e^2V_{\text{bias}}}\sqrt{\frac{\hbar}{2M\omega}}g\frac{\text{d}}{\text{d}z}\left(\frac{a^2}{\alpha^2+\beta^2}I(z)\right)$$

where a is the energy of the active electronic state in the molecule. α and β are the electronic couplings of the molecule with its two driving nano-electrodes (here the surface and the end tip apex), $I(z)$ is the driving current intensity depending on the z coordinate along the potential surface conformation change reaction path, M is the mass of the switched group, ω is its vibration frequency in its stable conformation potential well, and γ_{in} is the efficiency of the inelastic effect in this semi-classical approximation. This $F_{\text{in}}(I, z)$ force was added to the Newtonian equation of motion of the acetophenone molecule along its conformation change reaction path presented in Fig. 1 and calculated as described in Sect. 4. For simplicity, the possible conformation change of the chemisorbed acetophenone on Si(100) was restricted to a motion in one dimension z along this reaction path. Notice that in this model, only the ground-state potential energy surface was explored to get a and ω. We know how to go further and to implement a full quantum version of this through molecule inelastic current effect [7]. However, this fully quantum approach is actually too computing demanding because it requires the knowledge of the complete quantum vibration and rotation spectra (at least for their low-lying parts) of the surface molecule-latch.

From its stable planar conformation toward its standing up less stable conformation, the conformation change of an acetophenone is triggered by the tunnelling current intensity passing through this molecule. The corresponding molecular

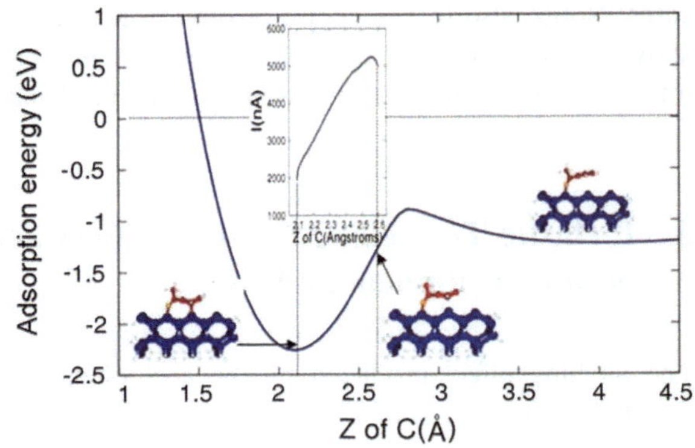

Fig. 12 A comparison between the conformation change reactive path found using the ASED+ optimization and the variation of the tunnelling current through this acetophenone molecule when the molecule is following this conformation reaction path

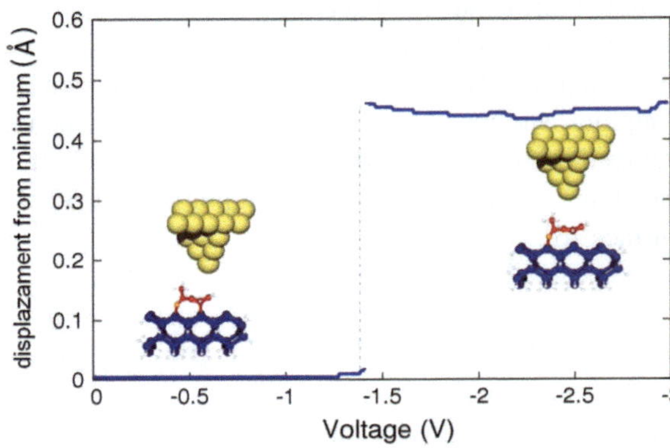

Fig. 13 Acetophenone conformation change triggered by the tip apex to surface tunnel current. The tip bias voltage must be larger than the Si(100) bulk gap to be large enough to be in competition with the acetophenone phenyl butterfly chemisorption

tunnel junction must reach a configuration to maximize the tunnel current intensity, keeping the molecule in the minimum possible conformation energy. This is illustrated in Fig. 12 where the ground-state potential energy curve of the acetophenone along the z reaction path is presented and compared with the $I(z)$ corresponding tunnelling current intensity variation.

The comparison between the two curves indicates that the initial and final conformations of the molecule are, respectively, $z = 2.1$ A and $z = 2.6$ A. The current intensity changes from 2 to 5.3 μA along the reaction path found for the tip to surface height selected for the calculations. Here, z is a reduced distance which is essentially the distance between the central carbon atom of the butterfly-like phenyl and the corresponding Si surface dimer.

These two curves were used as inputs for the dynamic Newton equation solver with the inelastic driving force expression given above to follow the acetophenone conformation change as a function of the STM tunnel junction bias voltage. As presented in Fig. 13 and after a bias voltage variation where the current intensity is not large enough to trigger a conformation change, there is a clear voltage threshold where the current driving force succeeds to break the bonding between the end phenyl and its surface DB dimer and to maintain this end phenyl in "suspension" above the DB dimer. This distance is enough to activate the conductance switching as presented above (see also Sect. 4.3 for more discussion about this point).

This inelastic semi-classical model demonstrates that even with a very stable ON state with the end phenyl of the acetophenone molecule well adsorbed on a DB dimer, it is possible to move up this end phenyl. This switching effect is not really a latching since the OFF state is maintained only at the expense of keeping active the tunnelling current drive. In the next section, a full optimization of this switching reaction path is presented to design better molecule switches for a true single molecule-latch effect.

4 Better Longer-Arm Flip-Flop Molecule-Latches

To better optimize molecule-latches and at the same time keep the idea of a flip-flop effect, we have implanted the nudged elastic band (NEB) algorithm [8] in our ASED+ code to have a better exploration of the ground-state reaction path of a molecular switching on Si(100) surface. This was applied to screen the surface switching characteristics of a series of arm-elongated acetophenone molecules to determine the best chemical structure of a molecule-latch for this series of molecules.

4.1 The NEB Reactive Path Search Implementation in ASED+

To calculate the minimum energy path (MEP) for a switching event between two possible configurations of a molecule-latch on a semi-conducting surface, the nudged elastic band (NEB) method introduced by D. Sheppard et al. in 2008 [8] was implemented in our ASED+ code [9]. The objective was to find molecules which are better than acetophenone to hold two conformations without the need of an inelastic tunnelling current flow to maintain each of them. We have selected the semi-empirical ASED+ code to be able to optimize the complete atomic scale circuit that is the molecule switch, the surface DB atomic wires and the contacting pads.

For the NEB method and as presented in Fig. 14, all the points along the MEP have the property of being minimums in the perpendicular direction to the path. To calculate the MEP, the minimum of energy for the initial and final states must be found first. The NEB algorithm starts with an initial guess of the MEP with N points (images) along it. This was obtained with a linear interpolation of the geometries from the starting point to the final one. In the algorithm, the images are connected by mechanical springs to ensure equal spacing along the reaction path. The forces perpendicular to the MEP plus the forces of the springs are relaxed. When these

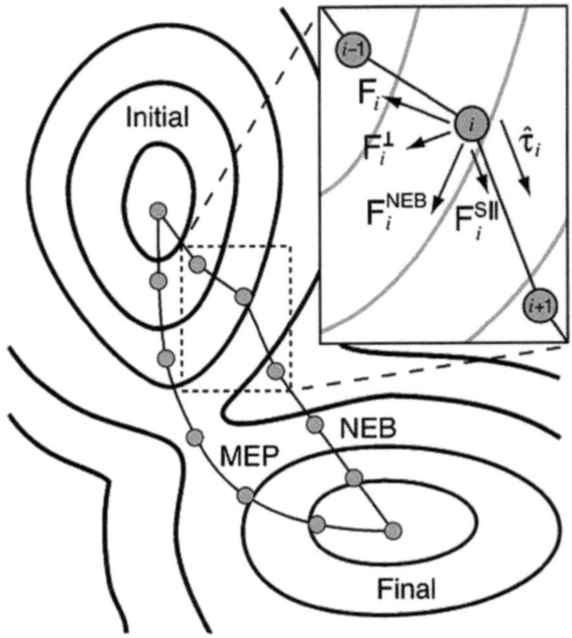

Fig. 14 Scheme of the nudged elastic band (NEB) method. The forces of each point in the NEB algorithm $\mathbf{F}_i^{NEB}$ are the sum of the spring force $\mathbf{F}_i^{S||}$ and the perpendicular force $\mathbf{F}_i^{\perp}$ to the tangent τ_i [8]

forces are equal to zero, then the MEP is found. The energy barrier for the switch is the one obtained as the energy to overcome the saddle point of the MEP.

The geometries of each conformation point along the MEP determine the switching mechanisms. The convergence of the NEB algorithm toward the MEP depends on the spring constant between the springs. Values between 1.0 and 0.1 eV/A^2 can be chosen for a faster convergence of the algorithm.

As an example, the MEP of the acetophenone molecule-latch chemisorbed on a Si(100) (2×1) surface as presented in Fig. 1 was calculated by this method (see also Fig. 16). The most stable configuration for the acetophenone is the "down" conformation with the molecule attached by the oxygen atom to the surface and a butterfly configuration for the end phenyl. The adsorption energy for this configuration is 2.2 eV. The UP conformation corresponds to a configuration where the molecule is only attached by its oxygen. The adsorption energy for this configuration is 1.25 eV. A starting point for the MEP was created using a linear interpolation of 20 points between the initial and the final state. When the algorithm converged and in order to obtain a smooth reaction path curve, the MEP was interpolated to 40 points and ran again. The result is presented in Fig. 16 with an energy profile of 40 points along the MEP. The energy barrier to pass from the UP to the DOWN configuration is 0.35 eV. For the reverse, an energy barrier of 1.3 eV is obtained. The convergence of the NEB algorithm with 20 initial points and 40 points at the end needed around 72-h total of CPU with ASED+. This technique was applied to the study all the other molecule switch presented below.

4.2 Testing Different Switching Arms for a Better Molecule-Latch

After having investigated the acetophenone molecule and to optimize the switching part of a flip-flop molecule-latch, we have explored the series of molecules presented in Fig. 15. The rationale behind the design of those molecules was to extend the switching arm, to preserve the end phenyl and the surface-anchoring pivot of the switch, and to push this pivot away from the DB dimer of the interferometer. In Fig. 15, the acetyl-biphenyl and the 4-acetylquaterphenyl have quite rigid arms, the 2- and 4-acetylalkylphenyl molecules are equipped with more flexible arms still terminated by a phenyl.

Our new ASED+ NEB code was applied to each of those molecules to precisely determine the switching reaction path on their ground-state potential energy surface while chemisorbed by the end pivot to a single dimer of the Si(100) (2×1) surface. As presented in Fig. 16, only 4-acetylalkylphenyl turns out to fulfill all the above-required conditions with also a very large switching barrier height between the 2 stable surface conformations.

4.3 Holding or Switching Over the Barrier?

From a surface mechanical switching point of view, we have determined that the 4-acetylalkylphenyl molecule is the one having the largest barrier height separation between its non-chemisorb and its chemisorb butterfly phenyl like conformations on the Si(100) (2×1) surface. At the same time, this molecule is showing almost the same adsorption energy for its 2 conformations. Therefore, we have calculated the *T*(*E*) spectra and the *I*–*V* curves of this 4-acetylalkylphenyl molecule-latch acting on a surface interferometer using the 2-ESQC technique including again the Si(100)H surface leakage current but using only a slab of 5 Si layer below the surface.

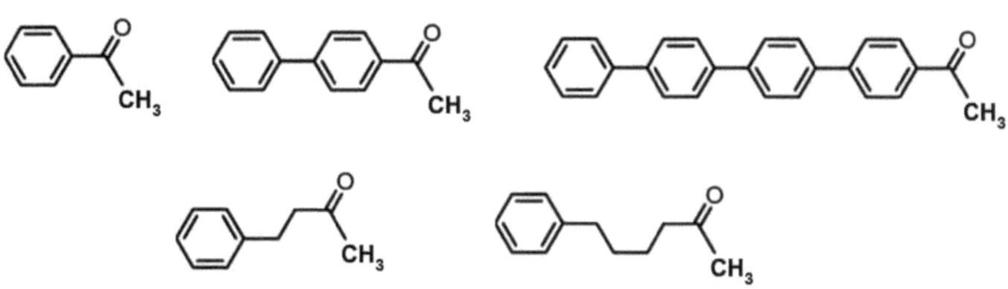

Fig. 15 Starting from the acetophenone molecule (*top left*), the different extension of the switching arm explored to control the chemisorption of the end phenyl: the acetyl-biphenyl, the 4-acetylquaterphenyl and the more floppy 2- and 4-acetylalkylphenyl molecules

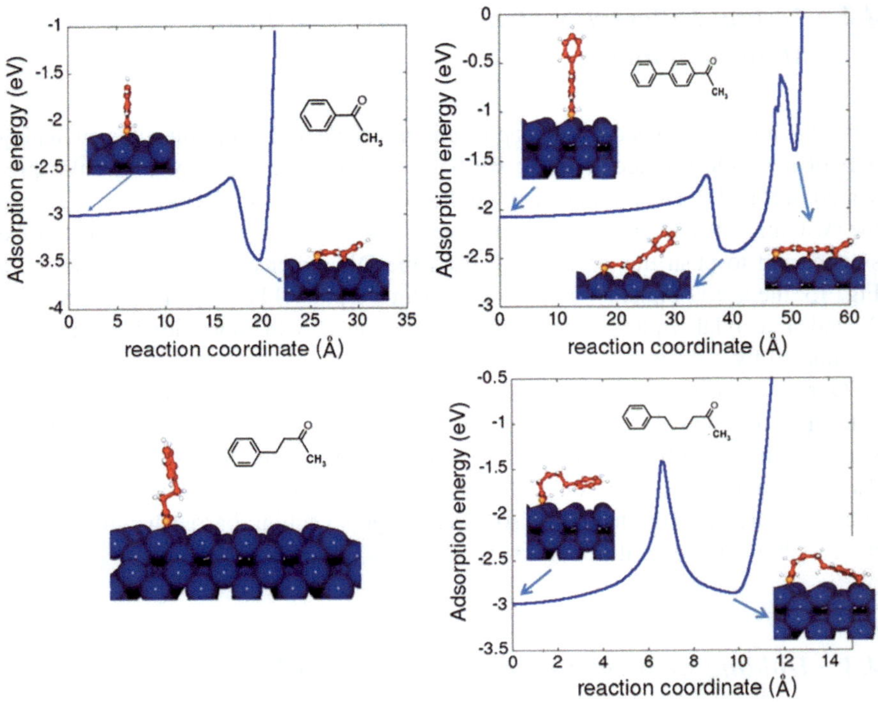

Fig. 16 Calculated minimum reaction path over the ground-state potential energy for four of the molecules presented in Fig. 15. The switching barrier heights are: 0.39 eV for acetophenone, 0.41 eV for acetyl-biphenyl, about the same for the long 4-acetylquaterphenyl (but with a lot of small minima on the phenyls chemisorption path), too short to bind by its phenyl for 2-acetylalkylphenyl, and 1.57 eV for 4-acetylalkylphenyl

A presented in Fig. 17 and using a very simple atomic scale surface arrangement for the molecule-latch, we found that the atomic scale quantum interferometer effect is still present using also a single lateral DB dimer to the central short DB wire. We have also restored a large tunnelling current intensity for the ON state and also a large ON/OFF ratio. The electronic valence band of the perpendicular to the dimer row dangling bond wire is very well restored in the ON state by the binding of the end phenyl of this 4-acetylalkylphenyl molecule.

Since the switching arm of the 4-acetylalkylphenyl molecule is quite soft, we have followed in detail how the approach of the end phenyl to the dangling bond dimer is progressively modifying the destructive interference effect. It turns out that for a phenyl–DB dimer distance just below 0.33 nm, the destructive interference is

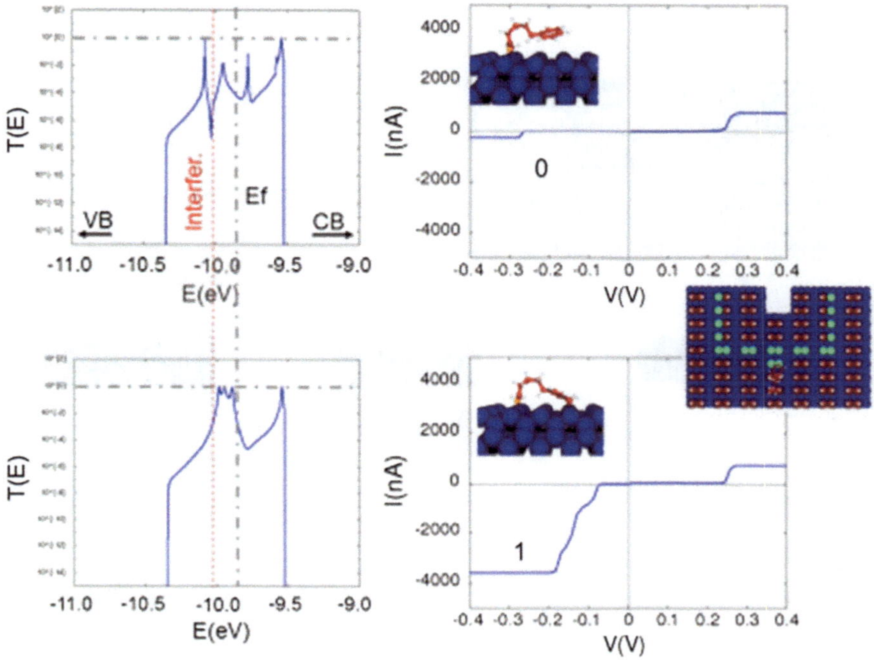

Fig. 17 *I–V* characteristics of the 4-acetylalkylphenyl molecule-latch with its corresponding *T*(*E*) spectra. The ON ("1") and OFF ("0") conformations are controlling the atomic scale interferometer, leading the tunnelling current through this device to be around 4 μA in the "ON" state and still a few nA in the "OFF" state. The binding of the 4-acetylalkylphenyl on the lateral dangling bond dimer of this interferometer is cancelling the very pronounced destructing interference effect indicated in red on the "0" OFF state *T*(*E*) spectrum

almost cancelled, indicating that it is not necessary to pass over the 1.57 eV switching barrier to control the tunnelling current. It is just enough to maintain a driving tunnelling current to approach the end phenyl enough from the dangling bond state without switching. Of course, this is not a molecule-latch any more. However, such a configuration can be very useful, for example, to obtain a power gain if the transition from A to C in the potential energy of Fig. 18 does not require much energy as compared to the output current of this atomic scale interferometer, which is varying by more than two orders of magnitude going from A to C. This effect reminds the C60 single-molecule transistor [10].

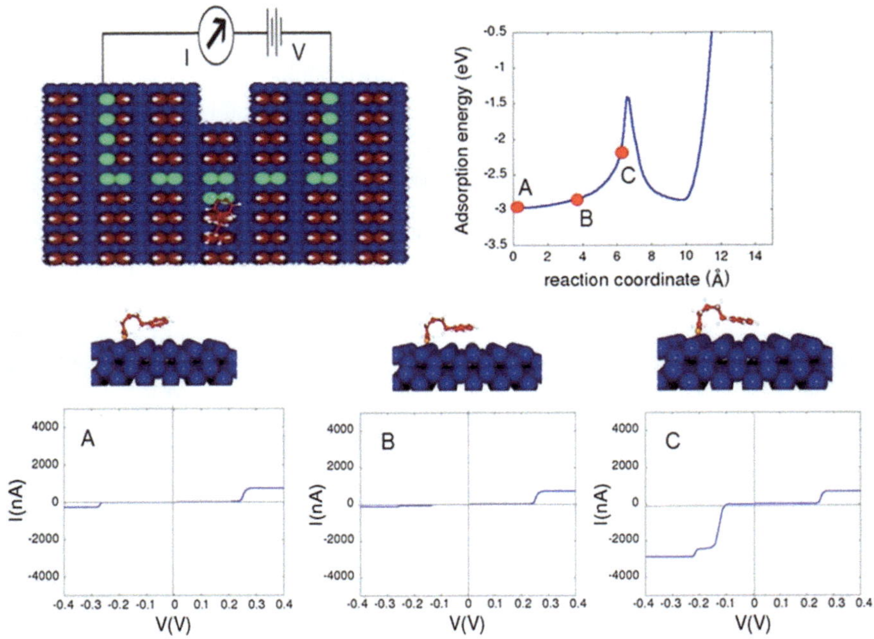

Fig. 18 Variation of the tunnelling current through the 4-acetylalkylphenyl molecule-latch without passing over the central 1.59 eV central energy barrier along the switching reaction path. The destructive interference effect is maintained in A and B, but cancelled out in C well before the switching saddle point

5 Conclusion

Starting with the description of three types of possible surface molecule switches that can allow the design of a molecule-latch to input data on an atomic scale circuit, the flip-flop switch was chosen. Mounted on a surface to control an electronic destructive interference effect, we have demonstrated that an atomic scale switch can be designed. A semi-classical model was presented to show how the inelastic tunnelling current effect can control the molecule-latch. The first studied acetophenone molecule-latch on Si(100)H does not present a true bistable behavior. We have further explored the acetophenone-like molecule to design a long arm switchable molecule with still the end phenyl required to control the atomic scale interferometer. A true bistable molecule-latch was designed with the required property. Those flip-flop surface molecule-latches need now to be used in more complex semi-classical and quantum circuits to demonstrate their digital data input capability.

References

1. Kawai, H., Ample, F., Qing, W., Yeo, Y.K., Saeys, M., Joachim, C.: Dangling bond logic gates on the Si(100)–(2x1)-H surface. J. Phys. Cond. Mat. **24**, 095011 (2012)
2. Schofield, S.R., Warschkow, O., Belcher, D.R., Rahnejat, K.A., Radny, M.W., Smith, P.V.: Phenyl attachment to Si(001) via STM manipulation of acetophenone. Phys. Chem. C **117**, 5736 (2013)
3. Neucheva, O.A., Ample, F., Joachim, C.: Mechanical conformation switching of a single pentacene molecule on Si(100)-2x1. J. Phys. Chem. C **49**, 117 (2013)
4. Godlewski, S., Kawai, H., Kolmer, M., Zuzak, R., Echevarren, A., Joachim, C., Szymonski, M., Saeys, M.: Single molecule rotational switch on a dangling bond dimer bearing. ACS Nano (2016, in press)
5. Nickel, A., Ohmann, R., Meyer, J., Grisolia, M., Joachim, C., Moresco, F., Cuniberti, G.: Moving nanostructures: pulse-induced positioning of supramolecular assemblies. ACS Nano **7**, 191 (2013)
6. Hliwa, M., Joachim, C.: Tunneling heating of a single Xe adsorbate. Phys. Rev. B **65**, 085406 (2002)
7. Monturet, S., Kepenenkian, M., Lorente, N., Joachim, C.: Inelastic transition selection rules. Chem. Phys. Lett. **573**, 1 (2013)
8. Sheppard, D., Terrell, R., Henkelman, G.: Optimization methods for finding minimum energy paths. J. Chem. Phys. **128**, 134106 (2008)
9. Ample, F., Joachim, C.: A semi-empirical study of polyacene molecules adsorbed on a Cu(110) surface. Surf. Sci. **600**, 3243 (2006)
10. Joachim, C., Gimzewski, J.K., Tang, H.: Physical Principles of the single C60 transistor effect. Phys. Rev. B, **58**, 16407 (1998)

Complex Atomic-Scale Surface Electronic Circuit's Simulator Including the Pads and the Supporting Surface

Francisco Ample, Omid Faizy and Christian Joachim

Abstract Complex atomic-scale circuits designed in classical, semi-classical and Quantum Hamiltonian Computing approaches using multi-atomic wires interconnects and Au(111) metallic nano-pads contacts discussed in Chapter "The Design of a Surface Atomic Scale Logic Gate with Molecular Latch Inputs" have been investigated to go forward for reaching compact and complex Boolean logic gates systems. The atomic wire electronic band gap, the tunneling intensity decay problems with the length of those atomic-scale interconnects and the practical size of the total atomic-scale circuit are discussed with the future perspective of developing an atomic-scale circuit simulator. Practical problems like the limitation in number of atoms of circuit builders and software limitations in terms of RAM memory and CPU computing time are also discussed for this objective.

1 Introduction

Following the design of elementary atomic-scale switches as, for example, presented in Chapter "The Design of a Surface Atomic Scale Logic Gate with Molecular Latch Inputs", the next step to design a complete surface atomic-scale Boolean logic circuit is to determine the planar implantation of the total circuit on the supporting surface. This applies for any classical [1], semi-classical [2] or

---

F. Ample
Institute of Materials Research and Engineering, 2 Fusionopolis Way, Innovis, #08-03, Singapore 138634, Singapore

O. Faizy · C. Joachim (✉)
GNS & MANA Satellite, CEMES-CNRS, 29 Rue J. Marvig, 31055 Toulouse Cedex, France
e-mail: joachim@cemes.fr

C. Joachim
International Center for Materials Nanoarchitectonics (WPI-MANA), National Institute for Materials Science (NIMS), 1-1 Namiki, Tsukuba, Ibaraki 305-0044, Japan

M. Kolmer and C. Joachim (eds.), *On-Surface Atomic Wires and Logic Gates*, Advances in Atom and Single Molecule Machines,
DOI 10.1007/978-3-319-51847-3_11

quantum [3] architectures. For this purpose, it is also necessary to bring metallic nano-pad interconnects to measure the logic output status of the circuit and the nano-pads for the inputs.

In this chapter, we use an example the classical, semi-classical and quantum designs of the simple example of an AND Boolean logic gate. The N-electrode elastic scattering quantum chemistry (N-ESQC) code is used to calculate the Boolean logical response of this simple circuit taking into account the atomic-scale circuit itself, the metallic nano-pads required and also the supporting surface and its bulk. N-ESQC is based on a full valence description of each atom of the system using a Slater basis set to get a good band structure of the supporting surface and of the metallic nano-pads [4]. No surface relaxation was considered in this chapter because it will bring many particular subcases and it is important first to draw a complete design before adapting it to the specific surface relaxation which is also depending on the bulk doping level under the surface (if any). A detail surface relaxation can be performed afterward to trouble check the robustness of a given atomic-scale implantation and feedback on the design again. In Sect. 2, the complete surface design of a molecule-latch atomic switch is presented taking also account the implantation of the latch controlling electrodes. In Sect. 3, the classical and semi-classical implementations of the AND gate on the Si(100)H surface are presented and in Sect. 3 its quantum version. In conclusion, we discuss the possible extension of this work toward the future construction of a true atomic-scale circuit simulator including the required molecule-latches, the complete supporting surface and the metallic nano-pads.

2 The Complete Surface Design of an Atomic Switch

Following the classical design for Boolean logic gates based on interconnecting switches together in series and in parallel along an electrical circuit, our first complete implantation concerns here, and on the Si(100)H surface, the use of atomic switches latched by a single acetophenone molecule each (see Chapter "Molecule-Latches in Atomic Scale Surface Logic Gates Constructed on Si(100)H"). As presented in Fig. 1, two kinds of atomic-scale switches can be constructed atom by atom depending on how the destructive interferences through the main dangling bond (DB) dimer atomic wire are configured relative to simple lateral dimers electron paths (Fig. 1). This is equivalent to a push or to a pull switch in terms of standard macroscopic switches as also indicated in Fig. 1. Both are working perfectly when simply interconnected to single DB atomic wires.

Following the standard rules of Boolean logic gate design, a series or a parallel classical (Ohms law) circuit connection between 2 of those atomic switches leads to the AND, OR, NAND and NOR Boolean logic function as presented in Fig. 2. We have not represented here the XOR and NXOR Boolean logic gates that can be designed associating in series or in parallel each of Fig. 1 switches. The reason is that according to the standard design of such a composite gate, gates need to be

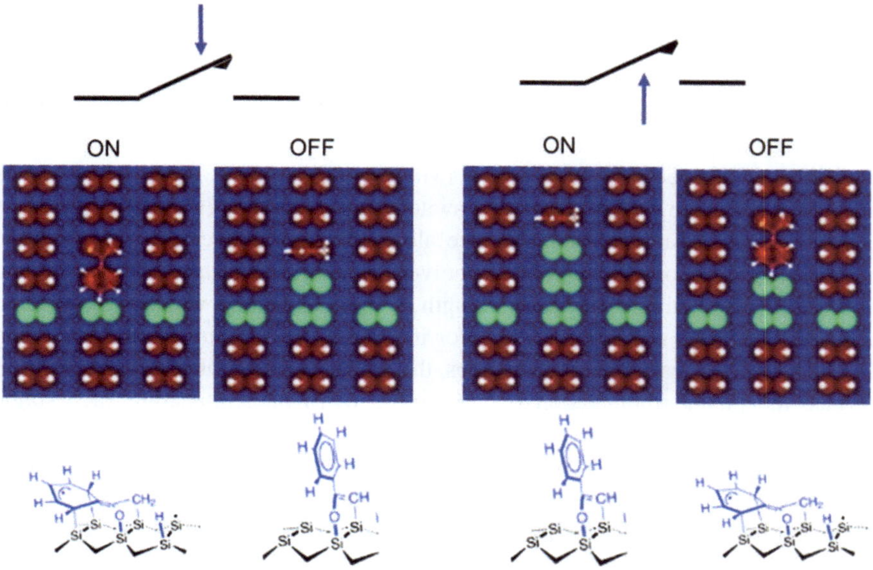

Fig. 1 Two types of electronic interference planar atomic-scale switches controlled by an acetophenone molecule-latch. Left for a push-like ON switch and right for push-like OFF switch (or equivalently a pull ON switch). Experimentally, the acetophenone is suppose here to be inelastically triggered by the tunneling current passing through the molecule when it is in a tunneling junction between the tip apex of an STM and the Si(100) surface

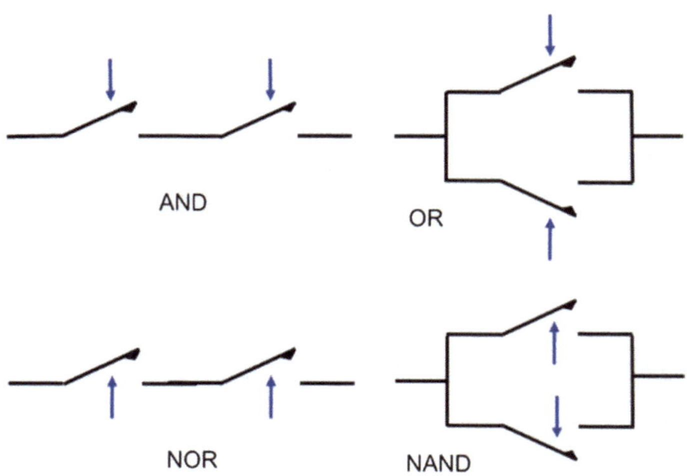

Fig. 2 Classical circuit implementation in series and in parallel of two identical switches coming from Fig. 1 and leading to the standard AND, OR, NAND and NOR Boolean logic functions

cascaded, meaning that inside the circuit, the output of an elementary gate must drive the input of the next output elementary gate. This is not actually possible with a molecule-latch.

Association in series or in parallel to construct the AND, OR, NAND and NOR requires an appreciation of the type of transport regime able to deliver those functions. In between the elementary switches, a Boltzmann transport regime is working well at the condition that the switches be positioned in space at distances larger than the mean free path distance along the interconnecting wires [1]. The ballistic and pseudo-ballistic can also be very good in some cases [2]. Notice that with its exponential decay with the length of interconnecting wires, the tunneling transport regime is not recommended for interconnects [5]. But before associating in series and/or in parallel those switches, the first step is to finalize on the Si(100)H surface the planar implantation of the atomic wires interconnecting each single atomic switch with its molecule-latch.

The first step is to design the best atomic wire interconnection for Fig. 1 switches. Hereafter we will only consider the case of the push-like switch (Fig. 1 left). The same optimization can also be performed for the pull-like switch (Fig. 1 right). Two types of interconnects can be constructed on the Si(100)H surface: parallel or perpendicular to the Si dimer rows. For such infinite DB wires and as presented in Fig. 3, the best interconnect configuration is the one with the DB wires perpendicular to the dimer rows. This is not coming from the band structure of those wires whose bandwidth is smaller for perpendicular case as compared to parallel. This is coming for a corner effect, i.e., the two atomic wires in the left configuration are connected perpendicular to the wire switch. This introduces two destructive interference effects at both interconnection corners due to through bond tunneling between the two next corner atoms at each corner.

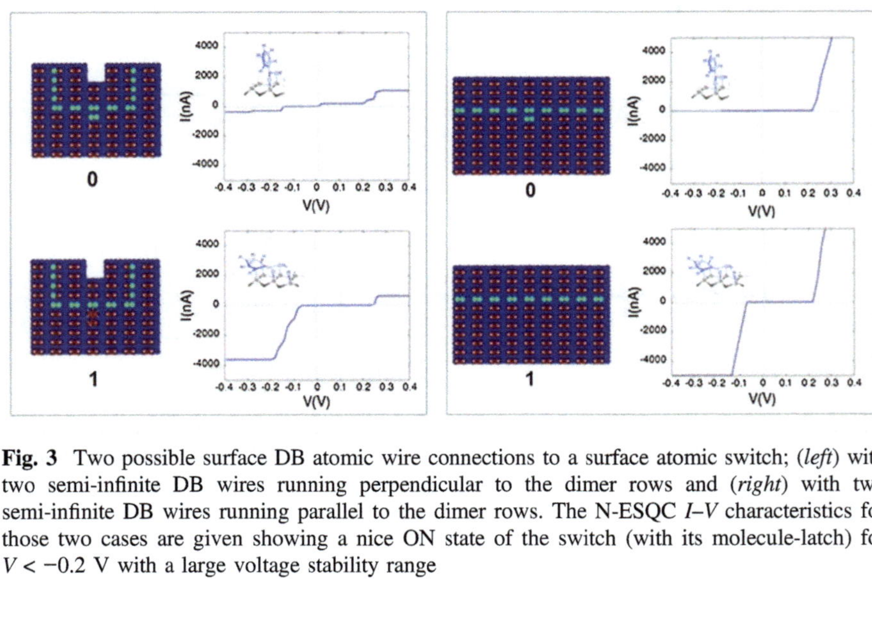

Fig. 3 Two possible surface DB atomic wire connections to a surface atomic switch; (*left*) with two semi-infinite DB wires running perpendicular to the dimer rows and (*right*) with two semi-infinite DB wires running parallel to the dimer rows. The N-ESQC *I*–*V* characteristics for those two cases are given showing a nice ON state of the switch (with its molecule-latch) for $V < -0.2$ V with a large voltage stability range

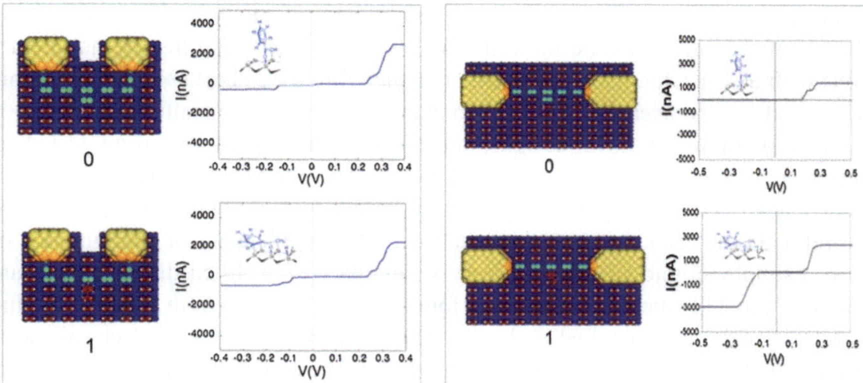

Fig. 4 Two Fig. 3 planar configurations equipped now with their respective planar metallic Au(111) interconnection nano-pads. Each *I–V* characteristic was obtained by calculating the multi-channel *T*(*E*) electronic transmission through the surface device and then calculating the current using the generalized Landauer–Buttiker formula. There are 5 Si layers to describe the central defect, and the Au(111) metallic nano-pads are 3 Au atomic layers in thickness and 7 Au atoms in width. There are supposed to be physisorbed on the Si(100)H surface electronically interacting with the nearest DB dimer through the orange like Au atoms whose 6 s Slater orbitals where parametrized with a double zeta basis set to recover the normal through space tunneling decay. The "0" and "1" configuration of the acetophenone molecule-latch is also indicated

We have also taken into account the destructive interferences effects to position on the surface the metallic interconnection nano-pads required to measure the ON/OFF status of the atomic switch. As presented in Fig. 4, N-ESQC was used to calculate the corresponding *I–V* characteristics with the metallic nano-pads positioned perpendicular or parallel to the switch with its molecule-latch. In the left planar interconnection configuration, the effect of interferences is enhanced simply because there is no gap in the electronic band structure of the Au(111) nano-pads. This leads to the capture of the complete destructive interference pattern. For the left interconnection configuration, the ON/OFF switch effect latched by the acetophenone is preserved. But the ON current is lower than in a pure DB wire interconnects because of the smaller electronic interaction between the top Au(111) metallic nano-pads and the DBs in contact with those nano-pads. This is also a nice illustration of an electronic constriction effect where the number of ballistic channels in the Au(111) metallic nano-pads is larger than the 2 channels of the atomic switch of the DB dimer wire.

The next step was to bring the two other metallic nano-pads to drive the molecule-latch switch in a planar configuration instead of using a top configuration STM tip apex as presented in Chapter "Molecule-Latches in Atomic Scale Surface Logic Gates Constructed on Si(100)H" of this book. It is unrealistic to directly bring two Au(111) metallic nano-pads in electronic interactions with the acetophenone molecule on the Si(100)H surface. First, they will be so closed that the tunneling current between those 2 nano-pads will pass through the surface and no inelastic effect will occur through the molecule. Second, the ON/OFF conformation change of the acetophenone molecule will be completely modified by its interactions with the metallic nano-pads. We have therefore used short surface DB atomic wires to drive the acetophenone switch starting from those metallic nano-pads.

After a lot of tentative designs, three surface configurations were tested as presented in Fig. 5. We have kept the best switch planar interconnection as discussed above, i.e., Fig. 4 left configurations. Figure 5 exemplifies the importance of a careful atomic-scale design and simulation of the complete planar configuration of even a simple atomic switch. In a very small space (7 $\times$ 14 Si dimers), there are now four Au(111) metallic nano-pads, four short DB wires in both crystallographic directions and a chemisorbed molecule. Therefore and in terms of electronic scattering calculations, the central defect representing the complete atomic switch has a very complex electronic structure requiring here N-ESQC calculations with $N = 4$ electrodes.

After N-ESQC had calculated the Fig. 5 configurations ($N = 4$ electrodes and 15 ballistic channels per Au(111) nano-pads), the surface implantation Fig. 5 A is best preserving all the switching characteristics even if a lot of tunneling current is passing through the Si(100)H surface between nano-pads 1 and 3. As presented in Fig. 6, the ON/OFF ratio of this switch is almost the same that the model Fig. 4 left case. The very large nano-pads 1–3 tunneling current is very important in terms of device switching because of the inelastic effect required to switch ON/OFF the molecule-latch. It is more a problem in terms of power consumption since even in the OFF state, this planar atomic switch will consume a lot of power.

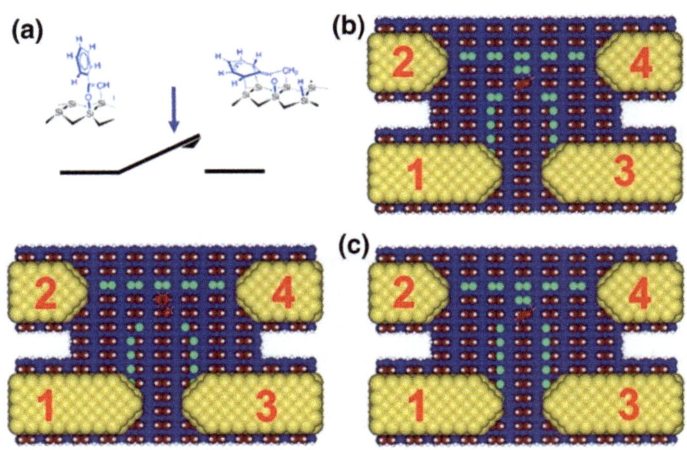

Fig. 5 Three short atomic wires configurations calculated to drive the acetophenone molecule-latch from the Au(111) metallic nano-pads. Metallic nano-pads 2 and 4 are the switch electrodes required to measure the ON/OF status of the switch. Metallic nano-pads 1 and 3 are driving the tunneling current through the central acetophenone molecule via the short DB wires. The different possible atomic-scale configuration of those two DB wires comes from the fact that through surface between those parallel DB wires must be avoided at max together with all the through surface tunneling leakage currents, for example, between the co-facial 1–2 electrodes

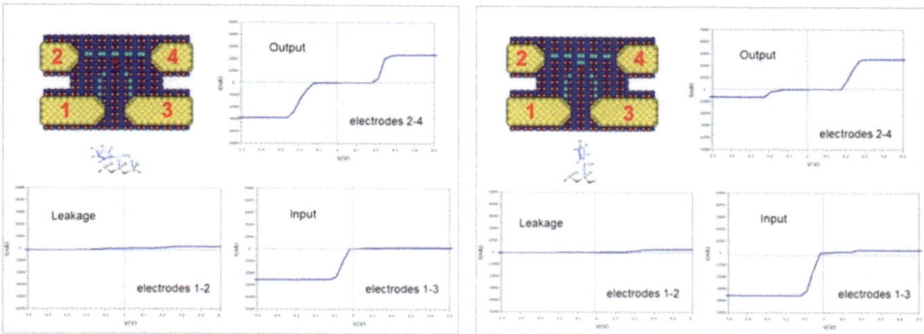

Fig. 6 All the characteristic tunneling currents through Fig. 5a model atomic switch controlled by a single acetophenone molecule. As in Fig. 5, the output current is calculated between 2 and 4, the driving current is between 1 and 3. The ON/OFF ratio is well preserved even if the OFF current is increasing due to the added driving DB wires in between 2 and 4. For all, the *I*–*V* curves voltage range between −0.5 and +0.5 V and current intensity between −5 and +5 μA

This is reminiscent to the old ECL transistor technology where a permanent current was passing through all the transistors any time to ensure an ultrafast response of the circuit. It may be possible to compensate this OFF current.

3 The Example of an AND Boolean Logic Gate

As recalled in Fig. 2, after having designing a nice switch and its complete planar surface implantation on Si(100)H, the next step is to associate those switches in series or in parallel. A first solution already explored with the single C60 transistor [1] is to interconnect them using standard wiring, i.e., with an electronic phase randomization between the switches, as presented in Fig. 7. For room-temperature operations, this would require a minimum distance of 10 nm between the interconnecting metallic pads, i.e., a distance equivalent to the electronic mean free path in gold at room temperature. This distance would have to be even larger at low temperature. Such classical interconnections are not very suitable simply because all the benefits of miniaturization would be lost by the length of those inter-switch interconnections.

A second solution is to work with metallic nano-interconnections whose lengths are below this mean free path as presented also in Fig. 7. Such nanoscale interconnection strategy is valid for the AND, OR, NAND and NOR gates recalled in Fig. 2, and the AND gate was just an illustrative example. In terms of number of atoms involved, those large atomic-scale circuits are at the moment out of the scope of the actual N-ESQC code version (see below in the Discussion section). There are 2640 Si surface atoms in Fig. 7 circuit surface implantation meaning a minimum of 52,800 atomic orbitals required to describe the electronic properties of this central circuit in terms of electronic scattering, but it may be not necessary to engage the

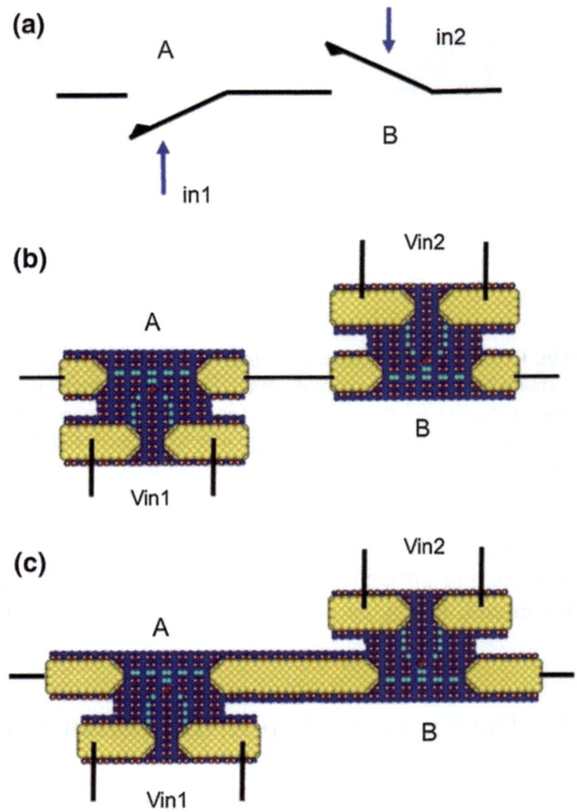

Fig. 7 Classical and mesoscopic ways of interconnecting in series two Fig. 6 atomic-scale switches to get an AND Boolean logic gate according to Fig. 2. **a** A simple reversal of Fig. 2 AND design. **b** An Ohmic electrical interconnect between 2 Fig. 6 atomic switches and **c** the A and B switches are directly interconnected through a metallic nanowire whose width is the same than the metallic Au(111) nano-pads

simulation of such a circuit since the intermediate metallic ballistic wire is using a lot of space.

To suppress the central long metallic nanowire interconnecting the A and B switches in Fig. 7, one solution is to use a simple DB wire as already explored in [2]. We have re-considered such an atomic-scale interconnect bringing now two acetophenone molecule-latches leading to a very small semi-classical AND gate. The two possible interconnections strategies discussed in Fig. 4 are also considered in Fig. 8. Here, the 2 Au(111) metallic nano-pad contacts positioned along the AND gate are again the best planar configuration for a functioning AND gate.

The problem with Fig. 8 configuration is that the central very short interconnection DB wire is a very bad wire with a very large electronic valence–conduction band gap. This means an exponential decay of the tunneling current with an increase in the length of this DB wire. This is the reason for keeping only one DB dimer in Fig. 8 right configuration instead of two in the left configuration to maintain a reasonable high-intensity output tunneling current. The consequence of this exponential decay of the tunneling current intensity with the length of the central atomic wire is that to introduce the 4 metallic nano-pads required to drive

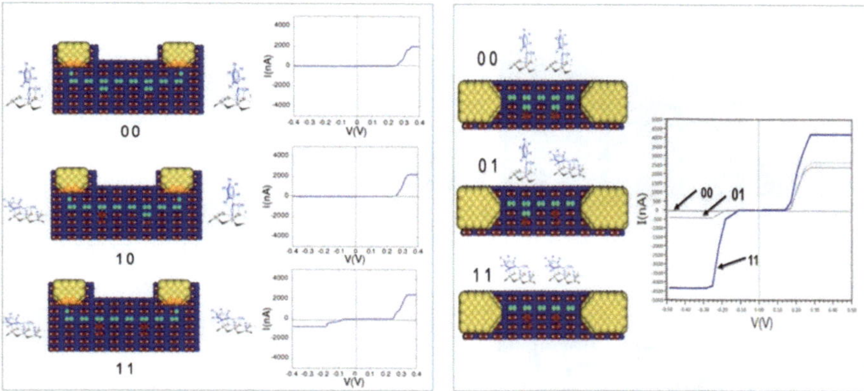

Fig. 8 Two interferometer-like atomic switches with their respective acetophenone switch interconnected in series through a very short DB atomic wire to construct an AND gate. For all, the *I*–*V* curves voltage range between −0.5 and +0.5 V and current intensity between −5 and +5 μA

the two acetophenone molecule-latches, one need to separate much more the switches than by using 2 DB dimers. This is not practicable.

4 The QHC AND Gate Complete Surface Implantation

Confronted to the design problem created by the nano-pads required to switch the molecule-latches, we have studied the planar implantation of a QHC DB AND logic gate starting from our QHC quantum graph expertise applied to DB atomic-scale circuit [3]. Here, N-ESQC was used up to $N = 5$ nano-electrodes. The starting point is our AND QHC Boolean logic gate playing with the shift in energy of the electronic states of a central DB dimer (Chapter "The Design of a Surface Atomic Scale Logic Gate with Molecular Latch Inputs" of this book). As presented in Fig. 9, this is working nicely with 2 DB interconnection wires used in parallel to the dimer rows.

In plugging the required 2 Au(111) metallic nano-pads to measure the logical output status of this QHC AND gate, we have followed the lessons learnt from the above semi-classical AND logic gate plugging and aligned those 2 pads along the central dangling bond dimer as presented in Fig. 10. The QHC AND gate is working perfectly, and the molecule-latch is playing its dangling bond saturation role with no problem. The problem came out now from the fact that with this aligned configuration of the metallic nano-pads, there is not so much space to bring the planar driving nano-pads for the two acetophenone switchings.

As presented in Fig. 11, we have simulated another planar interconnection configuration returning to a parallel to the dimer row metallic nano-pads alignment. Surprisingly and for this QHC AND gate, the AND truth table is also perfectly

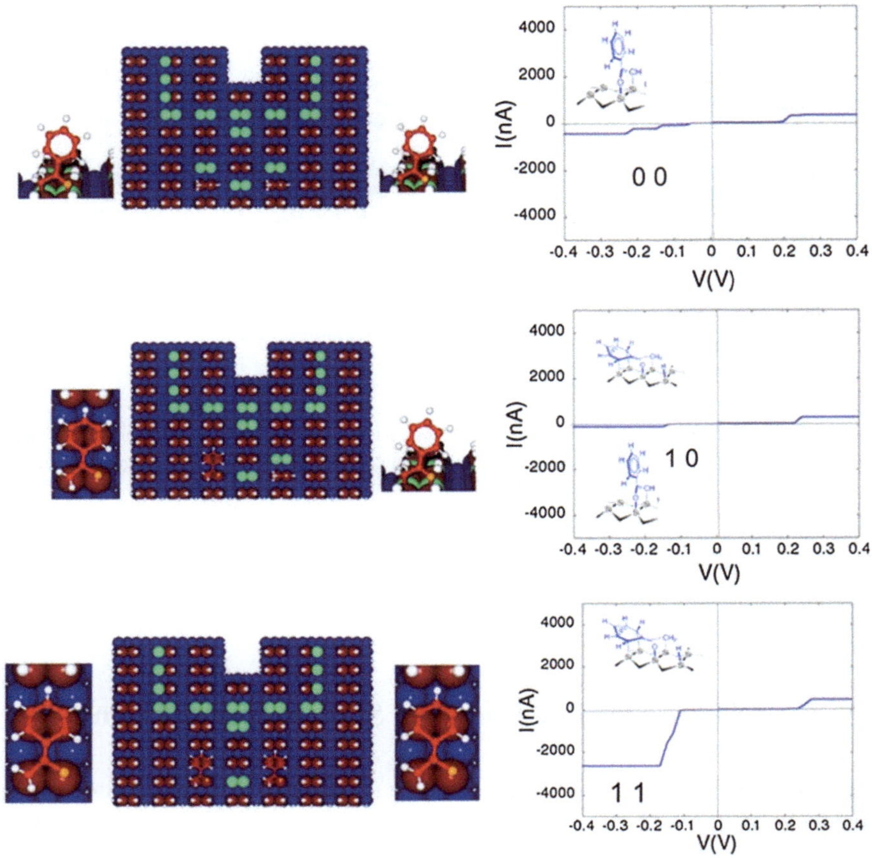

Fig. 9 A QHC AND gate constructed with 2 DB inputs triggered each by a single acetophenone molecule-latch. The AND gate is very stable for $V < -0.2$ V

respected. There is a decrease in the ON/OFF ratio. However, this is still manageable in terms of device performances. The interpretation is that with Fig. 11 interconnections, we have suppressed the corner effect with a direct electronic interactions between the end apex of each Au(111) facetted metallic nano-pad and the end of the QHC AND gate. Notice that in this N-ESQC for $N = 2$ calculations, we have not considered for simplification the molecule-latches to have a faster evaluation of the AND gate performance. The "(1,1)" tunneling current intensity is quite good with this new surface interconnection, and the QHC AND gate is very small.

Following Fig. 6 and with now the Fig. 11 QHC AND design, we have positioned on the Si(100)H surface the 3 complementary Au(111) facetted metallic nano-pads supposed to drive the acetophenone latching. This now a $N = 5$

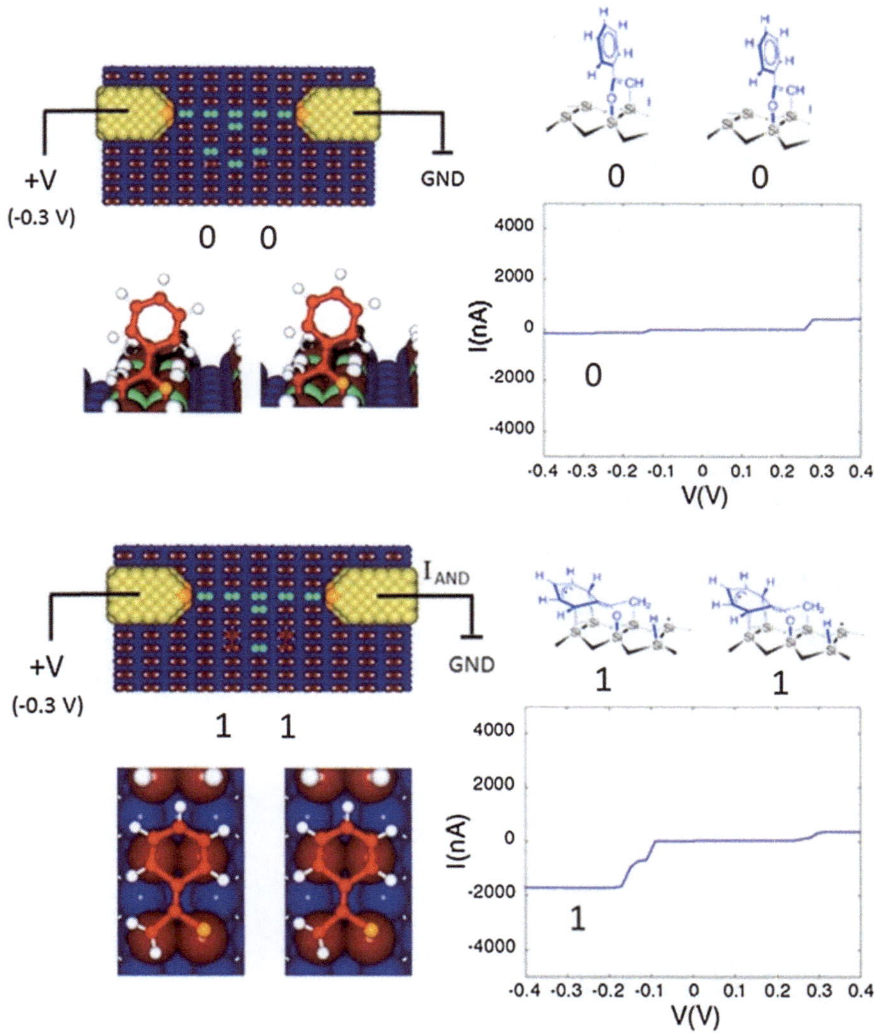

Fig. 10 A standard planar metallic nano-pads connection to Fig. 9 QHC AND gate. The AND truth table is perfectly preserved in this planar implantation

scattering problem for N-ESQC which demonstrates is nice stability in terms of the increase in the number of pads and scattering channels. As presented in Fig. 12, the 3 logical configurations were calculated with the surprised that now the AND logical output is shifted to the positive output voltage side. It may come out that the location of the nano-pads 4 and 5 very closed to the central DB edges of the QHC AND gate is shifting some of the electronic states of the gate.

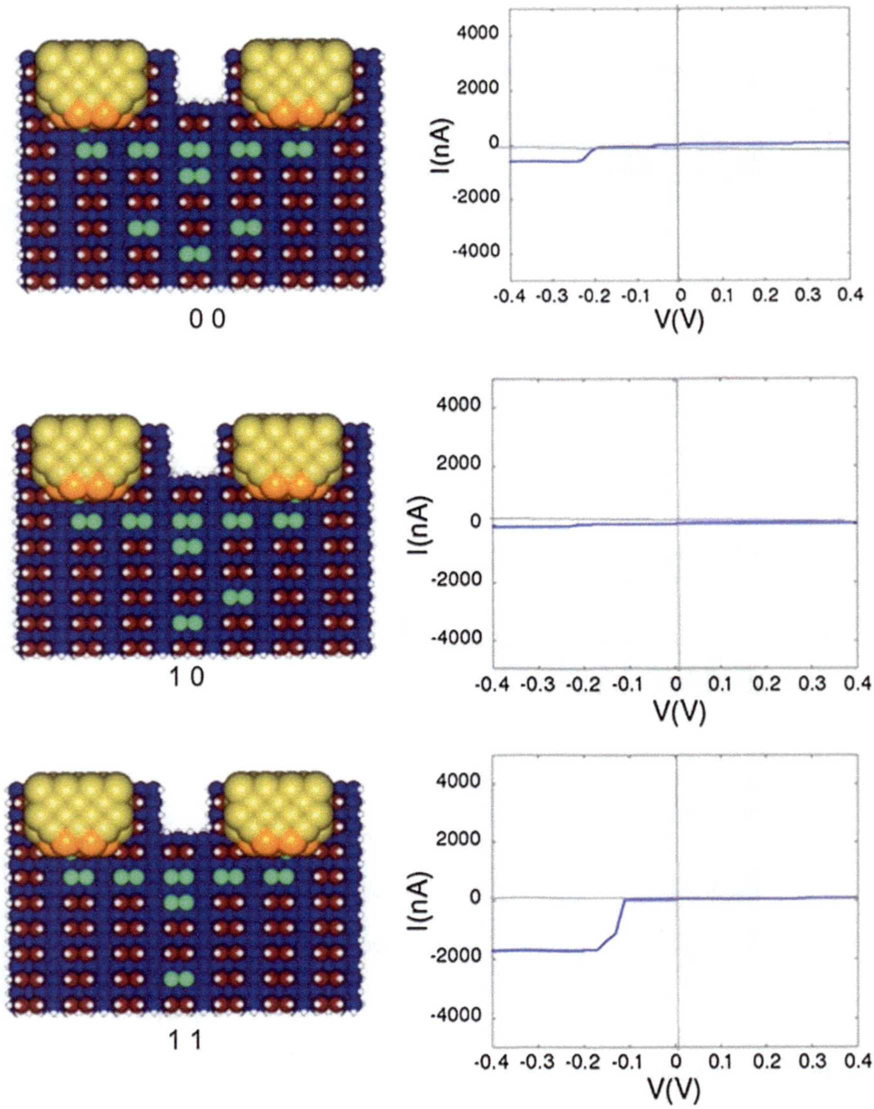

Fig. 11 A new implantation of a QHC AND gate with its interconnection electrode position parallel to the Si(100)H dimer rows. The AND truth table at −0.2 V is very well respected

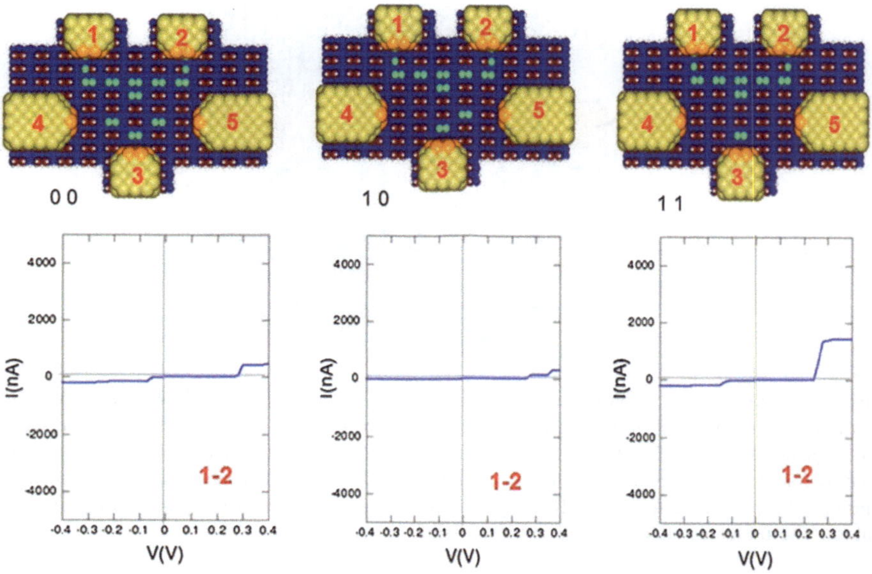

Fig. 12 Complete surface implementation on Si(100)H of a QHC AND gate with the output metallic nano-pads (1 and 2) and the driving nano-pads (3–4 and 3–5) for the molecule-latches of the logical input. The AND truth table is obtained for $V > +0.3$ V. For all, the I–V curves voltage range is between −0.4 and +0.4 V and current intensity between −5 and +5 μA

5 Discussion: Toward a Complete Atomic-Scale Circuit Simulator on a Surface

The transport calculations presented above are for very moderate size surface circuits but still taking into account the supporting surface electronic structure (and its bulk) and all the required metallic nano-pads interconnects adsorbed atop the Si(100)H surface. In those calculations, N-ESQC can accommodate up to about 2000 atoms for $N > 2$, i.e., a maximum of about 10,000 atomic orbitals (AO) to calculate the electronic transport characteristics of the top supported surface circuit in reasonable CPU time and RAM size. This 2000 atoms limitation is actually a true limitation for the future design of very complex atomic-scale circuits extending over hundreds of nanometers (see also Fig. 16).

An illustrative example of what an atomic-scale circuit simulator must be able to calculate in the future (at least in a non-time-dependent functioning regime) is given in Fig. 13. Here, we have represented a hypothetical semi-classical circuit resulting essentially from the superposition of elementary atomic-scale devices and wiring. This superposition is just for graphical illustration only using the simple hypothesis (not true) that the standard C. Shannon electrical superposition rules of logic gates are also working at the atomic scale.

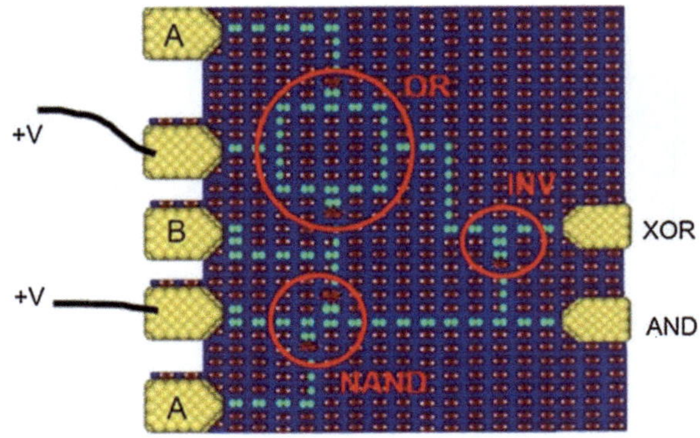

Fig. 13 A *top view* of an atomic-scale circuit digital ½ adder supposed to have been constructed atom by atom on an Si(100)H surface whose lateral size is 36 × 20 Si dimers. This represents a total of 8640 atoms (>35,000 atomic orbitals for a simple zeta full valence Slater basis set per atom) for the *central blue* part without counting the 7 Au metallic nano-pads each expending infinite. *Yellow* Au atoms, *white* H atoms, *red* Si surface atoms and *deep blue*: under surface Si atom. The molecule-latch switches are also visible in *red*

A very first practical problem for atomic-scale surface circuits as presented in Fig. 13 is the circuit visualization on a computer screen to anticipate the future needs for atomic-scale circuit builder. One will have to ensure the visualization of the circuit atom position on a screen and also their atom by atom extraction, for example, by clicking on the screen to generate the *x*, *y*, *z* input files required, for example, to run N-ESQC or one day the *N* electrodes version of DFT transport codes. On actual flat screens, the theoretical limit of visualization is around 10^6 atoms static and practical 10^5 atoms in a dynamic mode of representation that is with the required rotation of the surface and zoom in functionality.

The visual molecular dynamics (VMD) software was tested for a rectangular graphene supercell with 1,474,560 carbon atoms (see Fig. 14) using the CPK representation (named after the chemists Robert Corey, Linus Pauling and Walter Koltun) with atomic spheres and bonding lengths. The following information is used: number of spheres (atoms): 1,474,560 with a sphere scale of 1.0 and a sphere resolution of 12 and number of bonds: 2,616,194 with a bond radius of 0.3 and a bond resolution of 12.

A very important limitation of a builder is usually the time it takes to rotate the full atomic system. This is very important for molecular biology to have access to the different intramolecular cavity building up in a 3D protein structure. It is the same for an atomic-scale circuit where the supporting surface will have to be rotated, for example, for a good design of the back-interconnect electrodes up to the last surface layers.

For the graphene system of 1,474,560 atoms presented in Fig. 14, a rotation of 45° of the full sheet is taking about 40 s on a desk PC HP Z8000. Furthermore, the switching from an only-lines mode to a CPK mode takes about 37 s.

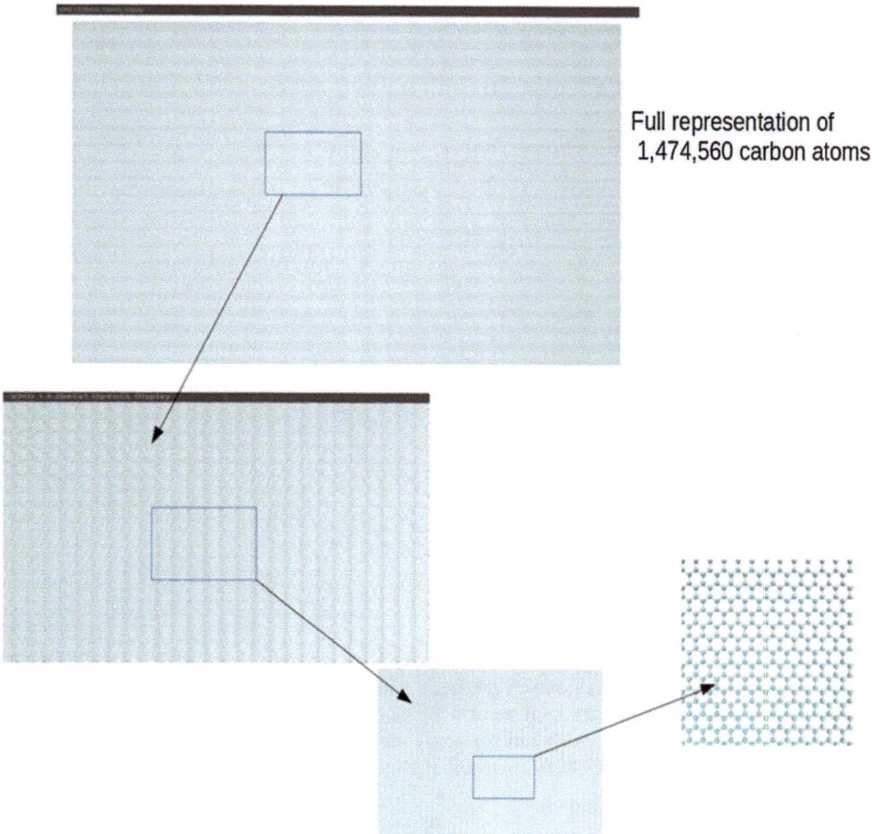

Fig. 14 Representation of a rectangular supercell of graphene with total number of 1,474,560 atoms with the progressive zoom into recover the C atoms. The memory size of the actual picture is ~4 GB, impossible to represent here to see its full resolution

We have also tested the *Molden* software often used by chemists. This software is not able to represent with a fast access a system of atoms with a number of atoms larger than 3000. The *PyMol* software was also tested for the same graphene system. We were able to represent such a large graphene system of 1,474,560 atoms on the screen. However, for systems larger than 400,000 atoms, we were not able to represent the bonds and can only represent the spheres.

Starting for a periodic and regular surface as presented in Figs. 13 and 14, it is important to able to zoom in fast and then to extract directly from the screen the atoms required to define the atomic-scale circuit. Then, a new *xyz* file must be generated to serve as an input file for a transport calculation code like N-ESQC. We have tested again the *PyMol* software, the only one to be able to deliver all those functionality as compared, for example, to VMD.

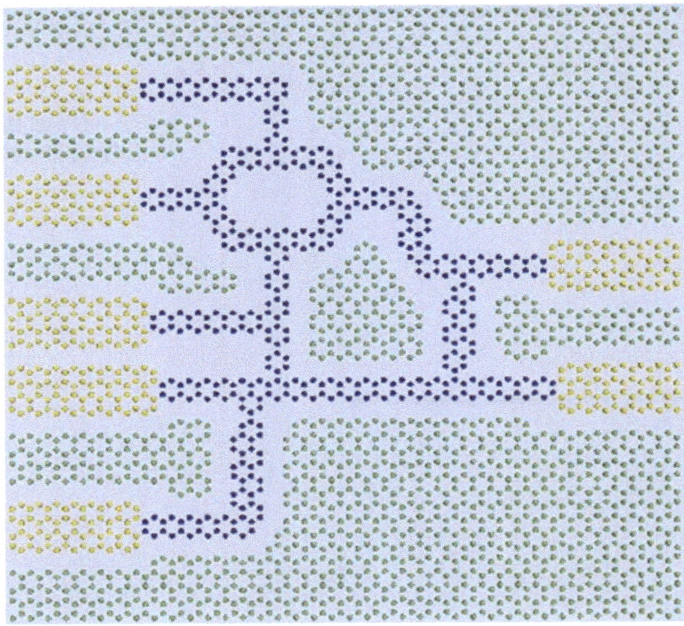

Fig. 15 An atomic-scale circuit digital ½ adder in analogy with Fig. 1 Si(100)H circuit. Here this circuit is constructed over a monolayer graphene with extracting the relevant carbon atoms by *PyMol* to construct a circuit with total number of 12,180 atomic orbitals. The interconnect is in *orange*, the passive surface in *green* and the active circuit in *dark blue*. Notice that this circuit is just a drawing to test the *PyMol* performances. It is not a functional circuit

We were able to remove and extract atoms one by one directly on the screen and by mouse operations. This is very convenient, and one can save the newly generated atomic system in a *xyz* file format compatible with transport calculation codes. As an example, we have made the graphene surface equivalent of the ½ adder circuit presented in Fig. 13 using only with 3045 carbon atoms (see Fig. 15). The N-ESQC code can immediately handle such a xyz file. Notice that *PyMol* is not VMD. For larger atomic-scale circuit, one will have to boost this *PyMol* software and run it on a very fast graphical card.

After exploring the graphical representation problem, we have explored on simple $N = 2$ ESQC example the limitations in CPU time and RAM memory and compare different calculation techniques. The maximum number of atomic orbital that ESQC can actually accommodate for $N = 2$ on large supercomputers is 64,744 AO leading to a very huge scattering matrix of 129,488 $\times$ 129,488 in Real*16 in ESQC. This scattering matrix is energy dependent and must be calculated hundreds of time to obtain the complete electronic transmission coefficient from any nano-pad to any nano-pad through the central atomic-scale circuit. More code developments are clearly required in the future to really develop an atomic-scale circuit simulator able to guide the experimental atom by atom construction of surface atomic-scale logic gate circuits. Such development can also find application

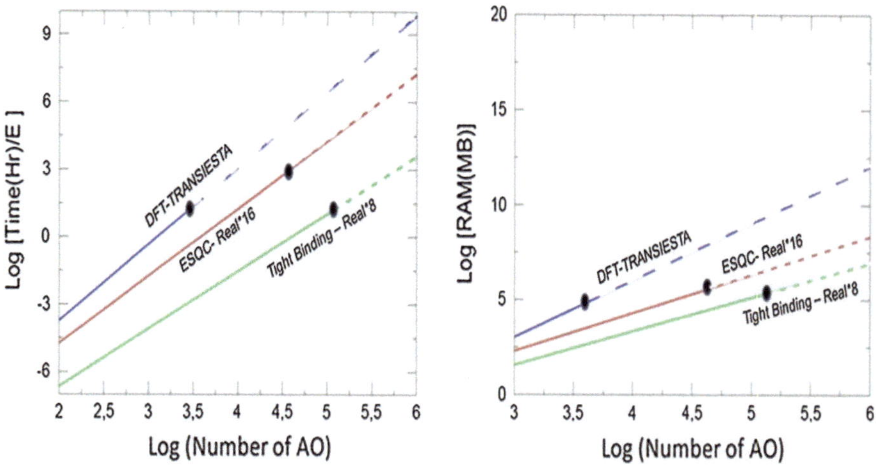

Fig. 16 ESQC atomic-scale circuit simulator roadmap for CPU and RAM as a function of the number of atomic orbitals in the circuit including the supporting surface. The 100,000 atoms target (i.e., approximatively 400,000 atomic orbitals) is around 5.5 on the abscissa. *Black point* is the reached calculation limits while writing this chapter

in the simulation of the next generation of CMOS transistor where the structure of the transistor will be described atom by atom including its grid. Here a Poisson solver will be certainly required running in a self-consistent way with N-ESQC.

Finally, a systematic study of the CPU and RAM power laws was performed to explore the limits of N-ESQC for $N = 2$ and provide the first atomic-scale simulator roadmap. As presented in Fig. 16, we have compared the semi-empirical $N = 2$ ESQC with the now well-used TRANSIESTA 2 electrodes transport code and with a very simple tight-binding scattering code with a simple one electronic state per atomic site and an orthogonal basis set. Both the CPU and RAM are clearly at the limits of the actual supercomputer resources.

References

1. Ami, S., Joachim, C.: Logic gates and memory cells based on single C60 electromechanical transistor. Nanotechnology **12**, 44 (2001)
2. Kawai, H., Ample, F., Qing, W., Yeo, Y.K., Saeys, M., Joachim, C.: Dangling bond logic gates on the Si(100)-(2x1)-H surface. J. Phys. Cond. Mat. **24**, 095011 (2012)
3. Kolmer, M., Zuzak, R., Godlewski, S., Szymonski, M., Dridi, G., Joachim, C.: Realization of a quantum hamiltonian computing Boolean logic gate on the Si(001):H surface. Nanoscale, **7**, 12325 (2015)
4. Lafferentz, L., Ample, F., Yu, H., Hercht, S., Joachim, C., Grill, L.: The conductance of a single conjugated polymer as a continuous function of its length. Science **323**, 1193 (2009)
5. Ample, F., Duchemin, I., Hliwa, M., Joachim, C.: Theoretical comparison between a single OR molecule gate and an atomic OR circuit logic gates interconnected on a Si(100)H surface. J. Phys. CM **23**, 125303 (2011)

MIX
Papier aus verantwortungsvollen Quellen
Paper from responsible sources
FSC® C105338

If you have any concerns about our products,
you can contact us on
ProductSafety@springernature.com

In case Publisher is established outside the EU,
the EU authorized representative is:
Springer Nature Customer Service Center GmbH
Europaplatz 3, 69115 Heidelberg, Germany

Printed by Libri Plureos GmbH
in Hamburg, Germany